# 2005
# Yearbook of
# Astronomy

# 2005 Yearbook of Astronomy

edited by
## Patrick Moore

co-editor
## John Mason

MACMILLAN

First published 2004 by Macmillan
an imprint of Pan Macmillan Ltd
20 New Wharf Road, London N1 9RR
Basingstoke and Oxford
Associated companies throughout the world
www.panmacmillan.com

ISBN 1 4050 4171 4

9 8 7 6 5 4 3 2 1

A CIP catalogue record for this book is available from
the British Library.

Typeset by Rowland Phototypesetting Ltd,
Bury St Edmunds, Suffolk
Printed and bound in Great Britain by
Mackays of Chatham plc, Chatham, Kent

# Contents

Part III
**Miscellaneous**

# Editors' Foreword

The *2005 Yearbook* follows the usual pattern, but with some important refinements; this year we have enlarged the star charts, drawn by Wil Tirion, for both Northern and Southern Hemispheres so that they will be more convenient to use, and we have also expanded the section summarizing the events for the year. The star charts to help you find Uranus and Neptune, which we introduced last year, seem to be popular and have been retained.

We have a varied selection of articles, many from our regular contributors and some from writers who are new to the *Yearbook* this year. As usual, we have done our best to give you a wide range, both of subject and of technical level. Gordon Taylor has, as always, provided the material for the monthly notes, and John Isles and Bob Argyle have provided the information on variable stars and double stars respectively.

PATRICK MOORE
JOHN MASON
Selsey, August 2004

# Preface

New readers will find that all the information in this *Yearbook* is given in diagrammatic or descriptive form; the positions of the planets may easily be found from the specially designed star charts, while the monthly notes describe the movements of the planets and give details of other astronomical phenomena visible in both the Northern and Southern hemispheres. Two sets of star charts are provided. The **Northern Charts** (pp. 17 to 41) are designed for use at latitude 52°N, but may be used without alteration throughout the British Isles, and (except in the case of eclipses and occultations) in other countries of similar northerly latitude. The **Southern Charts** (pp. 43 to 67) are drawn for latitude 35°S, and are suitable for use in South Africa, Australia and New Zealand, and other locations in approximately the same southerly latitude. The reader who needs more detailed information will find *Norton's Star Atlas* an invaluable guide, while more precise positions of the planets and their satellites, together with predictions of occultations, meteor showers and periodic comets, may be found in the *Handbook* of the British Astronomical Association. Readers will also find details of forthcoming events given in the American monthly magazine *Sky & Telescope* and the British periodicals *Astronomy Now* and *Astronomy and Space*.

**Important note**
The times given on the star charts and in the Monthly Notes are generally given as local times, using the 24-hour clock, the day beginning at midnight. All the dates and the times of a few events (e.g. eclipses) are given in Greenwich Mean Time (GMT), which is related to local time by the formula

Local Mean Time = GMT − west longitude

In practice, small differences in longitude are ignored, and the observer will use local clock time, which will be the appropriate Standard (or Zone) Time. As the formula indicates, places in west longitude will have a Standard Time slow on GMT, while places in east longitude will have a Standard Time fast on GMT. As examples we have:

**Standard Time in**

| | |
|---|---|
| New Zealand | GMT +12 hours |
| Victoria, NSW | GMT +10 hours |
| Western Australia | GMT + 8 hours |
| South Africa | GMT + 2 hours |
| British Isles | GMT |
| Eastern ST | GMT − 5 hours |
| Central ST | GMT − 6 hours, etc. |

If Summer Time is in use, the clocks will have been advanced by one hour, and this hour must be subtracted from the clock time to give Standard Time.

# Part I

# Monthly Charts and Astronomical Phenomena

# Notes on the Star Charts

The stars, together with the Sun, Moon and planets, seem to be set on the surface of the celestial sphere, which appears to rotate about the Earth from east to west. Since it is impossible to represent a curved surface accurately on a plane, any kind of star map is bound to contain some form of distortion.

Most of the monthly star charts which appear in the various journals and some national newspapers are drawn in circular form. This is perfectly accurate, but it can make the charts awkward to use. For the star charts in this volume, we have preferred to give two hemispherical maps for each month of the year, one showing the northern aspect of the sky and the other showing the southern aspect. Two sets of monthly charts are provided, one for observers in the Northern Hemisphere and one for those in the Southern Hemisphere.

Unfortunately, the constellations near the overhead point (the zenith) on these hemispherical charts can be rather distorted. This would be a serious drawback for precision charts, but what we have done is to give maps which are best suited to star recognition. We have also refrained from putting in too many stars, so that the main patterns stand out clearly. To help observers with any distortions near the zenith, and the lack of overlap between the charts of each pair, we have also included two circular maps, one showing all the constellations in the northern half of the sky, and one those in the southern half. Incidentally, there is a curious illusion that stars at an altitude of 60° or more are actually overhead, and beginners may often feel that they are leaning over backwards in trying to see them.

The charts show all stars down to the fourth magnitude, together with a number of fainter stars which are necessary to define the shapes of constellations. There is no standard system for representing the outlines of the constellations, and triangles and other simple figures have been used to give outlines which are easy to trace with the naked eye. The names of the constellations are given, together with the proper names of the brighter stars. The apparent magnitudes of the stars are indicated roughly by using different sizes of dot, the larger dots representing the brighter stars.

The two sets of star charts – one each for Northern and Southern Hemisphere observers – are similar in design. At each opening there is a single circular chart which shows all the constellations in that hemisphere of the sky. (These two charts are centred on the North and South Celestial Poles respectively.) Then there are twelve double-page spreads, showing the northern and southern aspects for each month of the year for observers in that hemisphere. In the **Northern Charts** (drawn for latitude 52°N) the left-hand chart of each spread shows the northern half of the sky (lettered 1N, 2N, 3N ... 12N), and the corresponding right-hand chart shows the southern half of the sky (lettered 1S, 2S, 3S ... 12S). The arrangement and lettering of the charts is exactly the same for the **Southern Charts** (drawn for latitude 35°S).

Because the sidereal day is shorter than the solar day, the stars appear to rise and set about four minutes earlier each day, and this amounts to two hours in a month. Hence, the twelve pairs of charts in each set are sufficient to give the appearance of the sky throughout the day at intervals of two hours, or at the same time of night at monthly intervals throughout the year. For example, charts 1N and 1S here are drawn for 23 hours on 6 January. The view will also be the same on 6 October at 05 hours; 6 November at 03 hours; 6 December at 01 hours and 6 February at 21 hours. The actual range of dates and times when the stars on the charts are visible is indicated on each page. Each pair of charts is numbered in bold type, and the number to be used for any given month and time may be found from the following table:

| Local Time | 18h | 20h | 22h | 0h | 2h | 4h | 6h |
|---|---|---|---|---|---|---|---|
| January | 11 | 12 | 1 | 2 | 3 | 4 | 5 |
| February | 12 | 1 | 2 | 3 | 4 | 5 | 6 |
| March | 1 | 2 | 3 | 4 | 5 | 6 | 7 |
| April | 2 | 3 | 4 | 5 | 6 | 7 | 8 |
| May | 3 | 4 | 5 | 6 | 7 | 8 | 9 |
| June | 4 | 5 | 6 | 7 | 8 | 9 | 10 |
| July | 5 | 6 | 7 | 8 | 9 | 10 | 11 |
| August | 6 | 7 | 8 | 9 | 10 | 11 | 12 |
| September | 7 | 8 | 9 | 10 | 11 | 12 | 1 |
| October | 8 | 9 | 10 | 11 | 12 | 1 | 2 |

| Local Time | 18ʰ | 20ʰ | 22ʰ | 0ʰ | 2ʰ | 4ʰ | 6ʰ |
|---|---|---|---|---|---|---|---|
| November | 9 | 10 | 11 | 12 | 1 | 2 | 3 |
| December | 10 | 11 | 12 | 1 | 2 | 3 | 4 |

On these charts, the ecliptic is drawn as a broken line on which longitude is marked every 10°. The positions of the planets are then easily found by reference to the table on p. 74. It will be noticed that on the **Southern Charts** the ecliptic may reach an altitude in excess of 62.5°on the star charts showing the northern aspect (5N to 9N). The continuations of the broken line will be found on the corresponding charts for the southern aspect (5S, 6S, 8S and 9S).

# Northern Star Charts

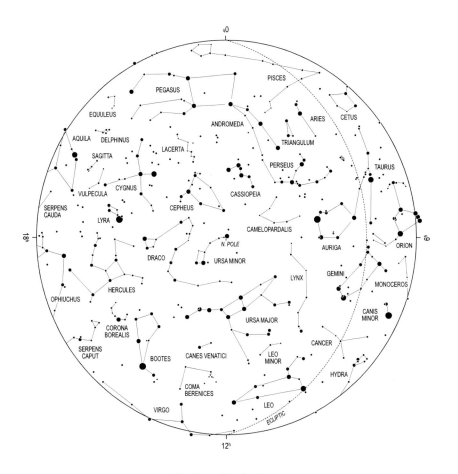

**Northern Hemisphere**

Note that the markers at 0ʰ, 6ʰ, 12ʰ and 18ʰ
indicate hours of Right Ascension.

# 1N

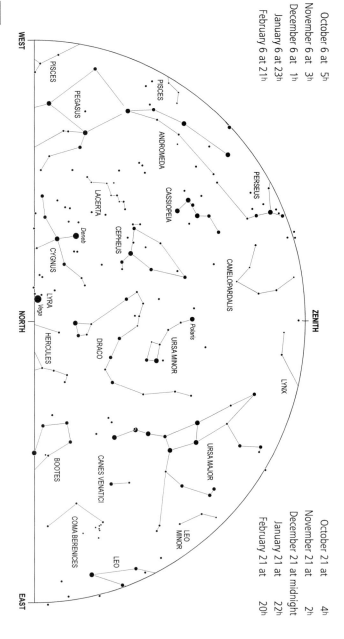

October 6 at 5h
November 6 at 3h
December 6 at 1h
January 6 at 23h
February 6 at 21h

October 21 at 4h
November 21 at 2h
December 21 at midnight
January 21 at 22h
February 21 at 20h

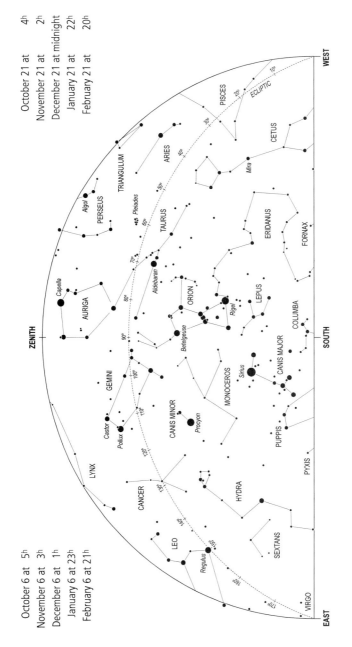

October 21 at 4ʰ
November 21 at 2ʰ
December 21 at midnight
January 21 at 22ʰ
February 21 at 20ʰ

October 6 at 5ʰ
November 6 at 3ʰ
December 6 at 1ʰ
January 6 at 23ʰ
February 6 at 21ʰ

WEST
PISCES
ECLIPTIC
CETUS
TRIANGULUM
ARIES
Mira
Algol
PERSEUS
Pleiades
TAURUS
ERIDANUS
FORNAX
Capella
AURIGA
Aldebaran
ORION
LEPUS
Rigel
COLUMBA
ZENITH
Betelgeuse
SOUTH
GEMINI
MONOCEROS
Sirius
CANIS MAJOR
Castor
Pollux
CANIS MINOR
Procyon
PUPPIS
LYNX
PYXIS
CANCER
HYDRA
LEO
SEXTANS
Regulus
VIRGO
EAST

## 2N

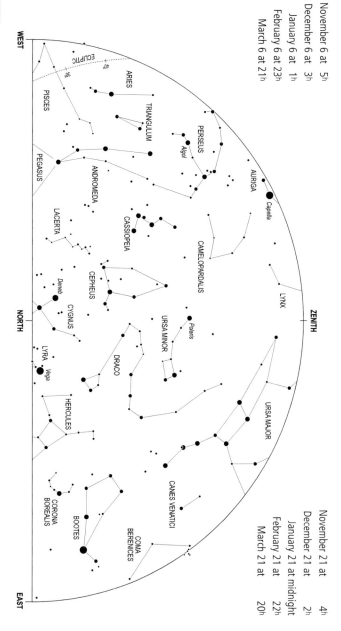

November 6 at 5h
December 6 at 3h
January 6 at 1h
February 6 at 23h
March 6 at 21h

November 21 at 4h
December 21 at 2h
January 21 at midnight
February 21 at 22h
March 21 at 20h

**2S**

November 21 at    4ʰ
December 21 at    2ʰ
January 21 at midnight
February 21 at    22ʰ
March 21 at       20ʰ

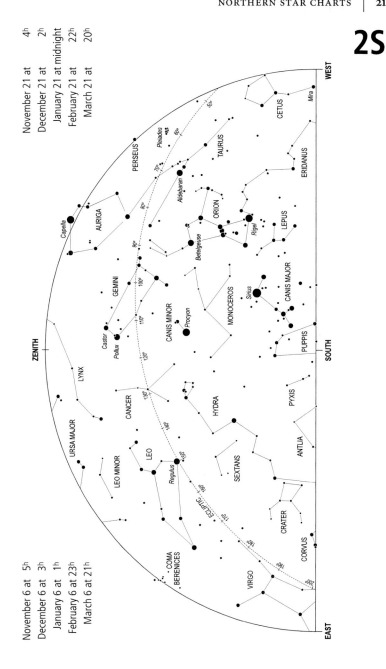

ZENITH

WEST

EAST

SOUTH

November 6 at   5ʰ
December 6 at   3ʰ
January 6 at    1ʰ
February 6 at   23ʰ
March 6 at      21ʰ

# 3N

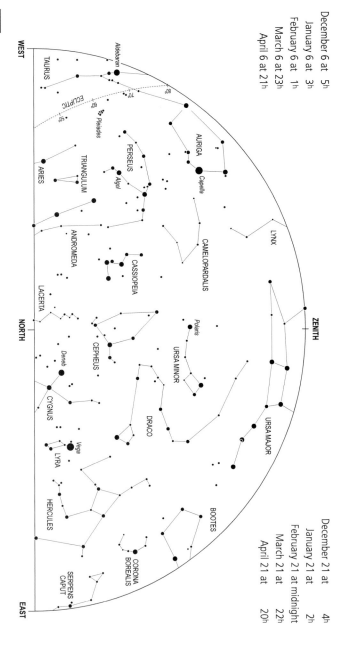

December 6 at 5ʰ
January 6 at 3ʰ
February 6 at 1ʰ
March 6 at 23ʰ
April 6 at 21ʰ

December 21 at 4ʰ
January 21 at 2ʰ
February 21 at midnight
March 21 at 22ʰ
April 21 at 20ʰ

WEST

TAURUS
Aldebaran
ECLIPTIC
80°
70°
60°
50°
Pleiades
AURIGA
Capella
PERSEUS
ARIES
TRIANGULUM
Algol
ANDROMEDA
CAMELOPARDALIS
LYNX
ZENITH
CASSIOPEIA
LACERTA
CEPHEUS
Polaris
URSA MINOR
NORTH
URSA MAJOR
Deneb
CYGNUS
DRACO
Vega
LYRA
HERCULES
BOOTES
CORONA BOREALIS
SERPENS CAPUT
EAST

**3S**

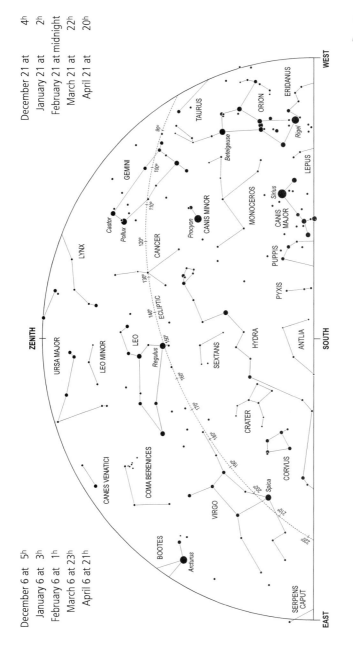

WEST

ERIDANUS

ORION

TAURUS

*Rigel*

*Betelgeuse*

GEMINI

LEPUS

*Castor*

*Pollux*

LYNX

CANIS MINOR

*Procyon*

MONOCEROS

*Sirius*

CANIS MAJOR

CANCER

PUPPIS

ZENITH

URSA MAJOR

LEO MINOR

LEO

ECLIPTIC

*Regulus*

SEXTANS

HYDRA

PYXIS

ANTLIA

SOUTH

CANES VENATICI

COMA BERENICES

CRATER

CORVUS

*Spica*

VIRGO

BOOTES

*Arcturus*

SERPENS CAPUT

EAST

**4N**

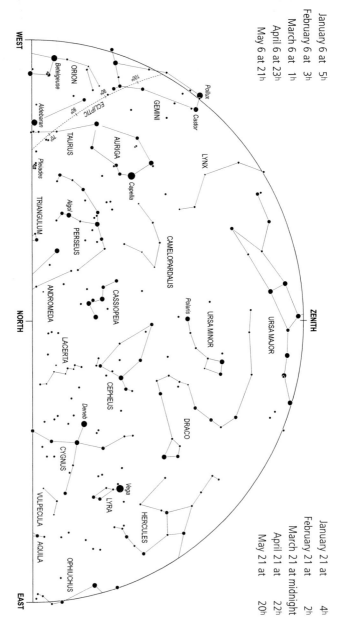

January 6 at 5ʰ
February 6 at 3ʰ
March 6 at 1ʰ
April 6 at 23ʰ
May 6 at 21ʰ

January 21 at 4ʰ
February 21 at 2ʰ
March 21 at midnight
April 21 at 22ʰ
May 21 at 20ʰ

WEST

EAST

NORTH

ZENITH

ORION
Betelgeuse
Aldebaran
GEMINI
Pollux
Castor
ECLIPTIC
TAURUS
AURIGA
Pleiades
Capella
LYNX
TRIANGULUM
Algol
PERSEUS
CAMELOPARDALIS
ANDROMEDA
CASSIOPEIA
Polaris
URSA MINOR
URSA MAJOR
LACERTA
CEPHEUS
DRACO
Deneb
CYGNUS
Vega
LYRA
HERCULES
VULPECULA
AQUILA
OPHIUCHUS

**4S**

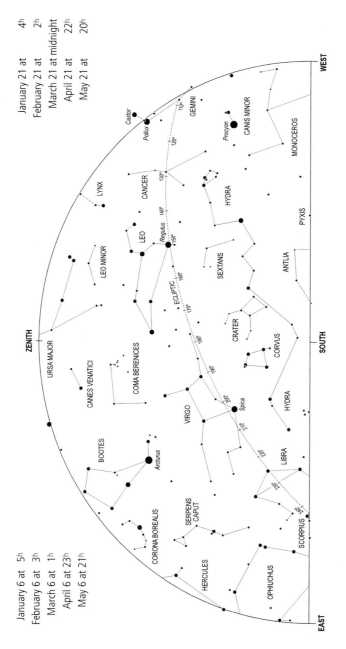

# 5N

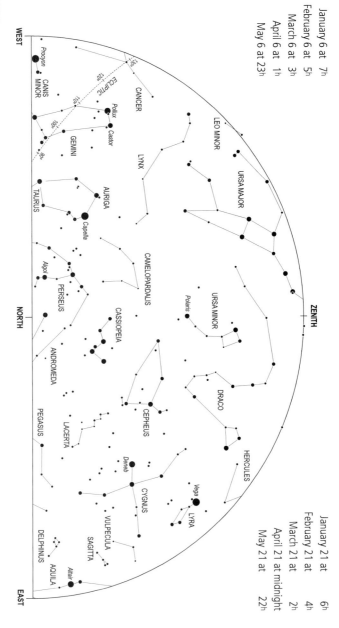

January 21 at 6h
February 21 at 4h
March 21 at 2h
April 21 at midnight
May 21 at 22h

**5S**

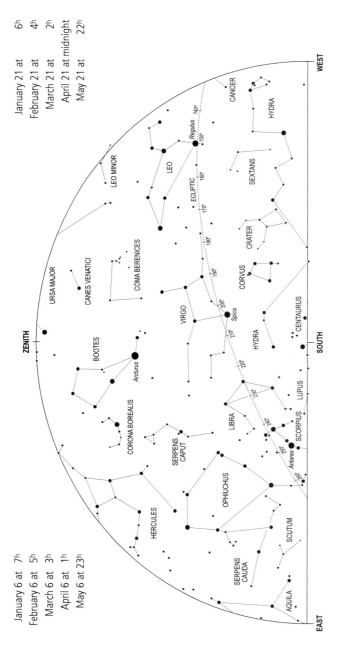

# 6N

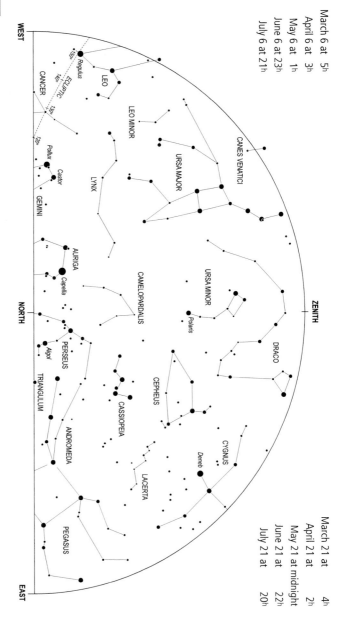

**6S**

March 21 at 4ʰ
April 21 at 2ʰ
May 21 at midnight
June 21 at 22ʰ
July 21 at 20ʰ

March 6 at 5ʰ
April 6 at 3ʰ
May 6 at 1ʰ
June 6 at 23ʰ
July 6 at 21ʰ

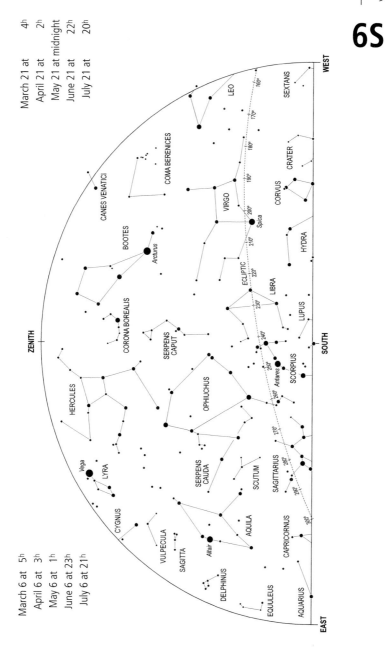

# 7N

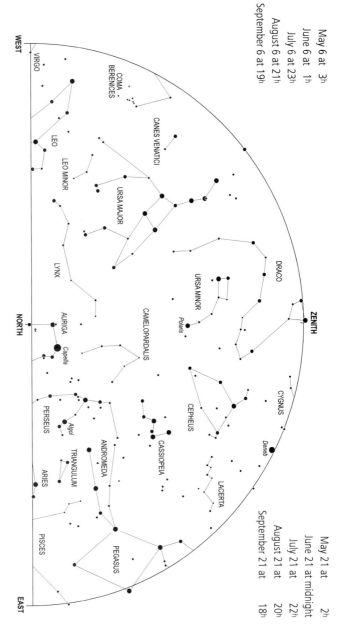

# 7S

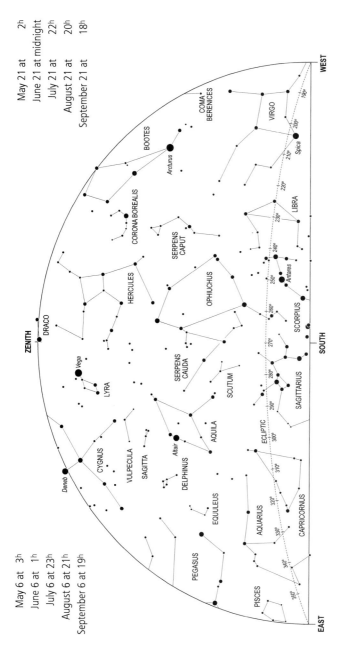

May 21 at 2ʰ
June 21 at midnight
July 21 at 22ʰ
August 21 at 20ʰ
September 21 at 18ʰ

May 6 at 3ʰ
June 6 at 1ʰ
July 6 at 23ʰ
August 6 at 21ʰ
September 6 at 19ʰ

ZENITH

WEST

EAST

SOUTH

COMA BERENICES
VIRGO
Spica
BOOTES
Arcturus
LIBRA
CORONA BOREALIS
SERPENS CAPUT
Antares
SCORPIUS
OPHIUCHUS
HERCULES
DRACO
Vega
LYRA
SERPENS CAUDA
SCUTUM
SAGITTARIUS
ECLIPTIC
VULPECULA
SAGITTA
CYGNUS
Altair
AQUILA
Deneb
DELPHINUS
CAPRICORNUS
EQUULEUS
AQUARIUS
PEGASUS
PISCES

190° 200° 210° 220° 230° 240° 250° 260° 270° 280° 290° 300° 310° 320° 330° 340° 350°

**8N**

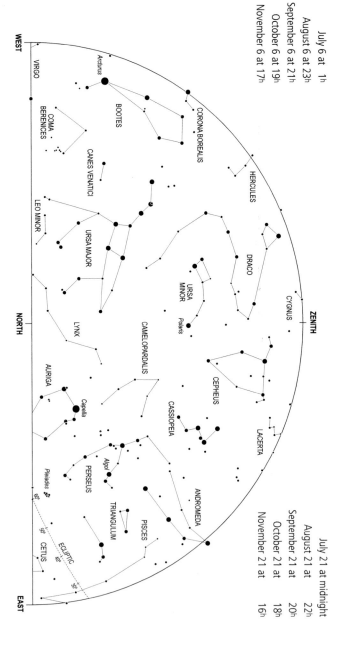

July 6 at 1h
August 6 at 23h
September 6 at 21h
October 6 at 19h
November 6 at 17h

WEST

VIRGO

Arcturus

BOOTES

COMA BERENICES

CORONA BOREALIS

HERCULES

CANES VENATICI

LEO MINOR

URSA MAJOR

DRACO

URSA MINOR
Polaris

CYGNUS

ZENITH

LYNX

CAMELOPARDALIS

CEPHEUS

AURIGA

Capella

CASSIOPEIA

LACERTA

Algol
PERSEUS

Pleiades

ANDROMEDA

60°

50°

TRIANGULUM

PISCES

ECLIPTIC
40°

CETUS

30°

NORTH

EAST

July 21 at midnight
August 21 at 22h
September 21 at 20h
October 21 at 18h
November 21 at 16h

**8S**

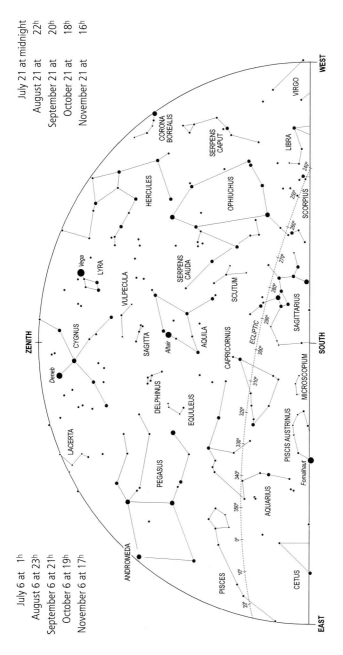

July 21 at midnight
August 21 at 22ʰ
September 21 at 20ʰ
October 21 at 18ʰ
November 21 at 16ʰ

July 6 at 1ʰ
August 6 at 23ʰ
September 6 at 21ʰ
October 6 at 19ʰ
November 6 at 17ʰ

ZENITH

WEST

SOUTH

EAST

VIRGO

CORONA
BOREALIS

SERPENS
CAPUT

LIBRA

HERCULES

OPHIUCHUS

SCORPIUS

240°

250°

260°

Vega
LYRA

VULPECULA

SERPENS
CAUDA

SCUTUM

270°

CYGNUS

Deneb

SAGITTA

Altair

AQUILA

CAPRICORNUS

SAGITTARIUS

280°

290°

ECLIPTIC

300°

DELPHINUS

EQUULEUS

310°

MICROSCOPIUM

LACERTA

320°

PISCIS AUSTRINUS

PEGASUS

330°

Fomalhaut

340°

ANDROMEDA

AQUARIUS

350°

0°

10°

PISCES

CETUS

20°

# 9N

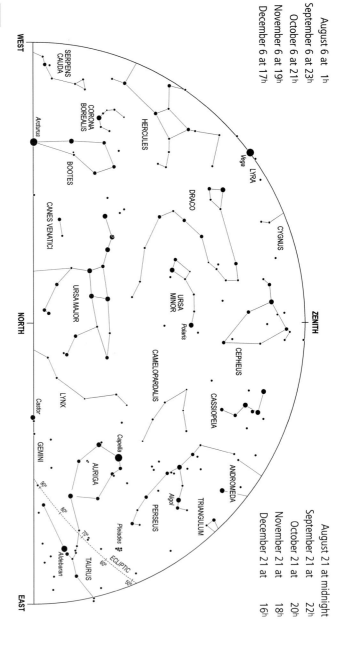

August 6 at 1h
September 6 at 23h
October 6 at 21h
November 6 at 19h
December 6 at 17h

August 21 at midnight
September 21 at 22h
October 21 at 20h
November 21 at 18h
December 21 at 16h

**9S**

August 21 at midnight
September 21 at 22ʰ
October 21 at 20ʰ
November 21 at 18ʰ
December 21 at 16ʰ

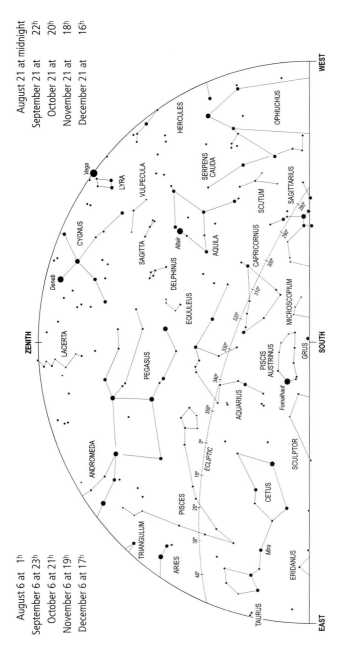

ZENITH

WEST

HERCULES
OPHIUCHUS
Vega
LYRA
VULPECULA
SERPENS CAUDA
CYGNUS
SAGITTA
Altair
AQUILA
SCUTUM
SAGITTARIUS
280°
Deneb
DELPHINUS
EQUULEUS
CAPRICORNUS
290°
300°
LACERTA
MICROSCOPIUM
310°
PEGASUS
320°
PISCIS AUSTRINUS
GRUS
330°
Fomalhaut
AQUARIUS
340°
SOUTH
ANDROMEDA
350°
0°
ECLIPTIC
10°
SCULPTOR
PISCES
20°
CETUS
TRIANGULUM
30°
Mira
ARIES
ERIDANUS
40°
TAURUS

EAST

August 6 at 1ʰ
September 6 at 23ʰ
October 6 at 21ʰ
November 6 at 19ʰ
December 6 at 17ʰ

# 10N

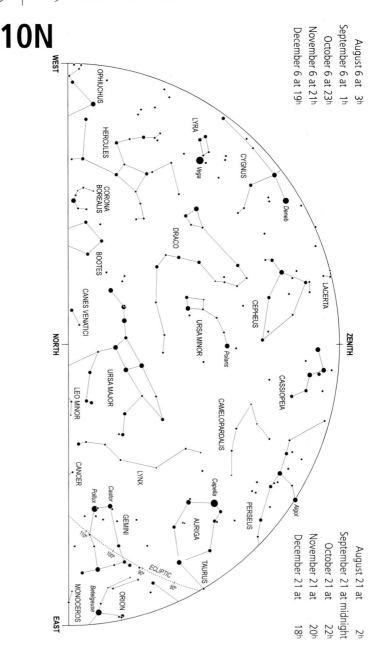

August 6 at 3h
September 6 at 1h
October 6 at 23h
November 6 at 21h
December 6 at 19h

August 21 at 2h
September 21 at midnight
October 21 at 22h
November 21 at 20h
December 21 at 18h

# 10S

August 21 at 2ʰ
September 21 at midnight
October 21 at 22ʰ
November 21 at 20ʰ
December 21 at 18ʰ

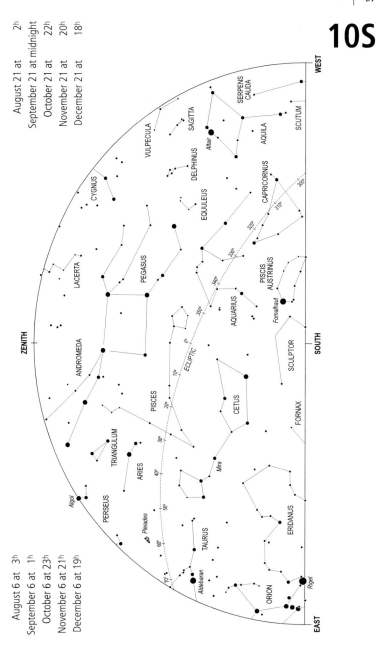

August 6 at 3ʰ
September 6 at 1ʰ
October 6 at 23ʰ
November 6 at 21ʰ
December 6 at 19ʰ

# 11N

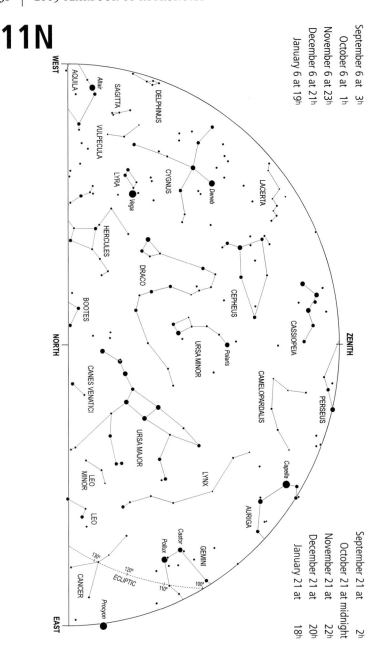

September 6 at 3h
October 6 at 1h
November 6 at 23h
December 6 at 21h
January 6 at 19h

September 21 at 2h
October 21 at midnight
November 21 at 22h
December 21 at 20h
January 21 at 18h

**11S**

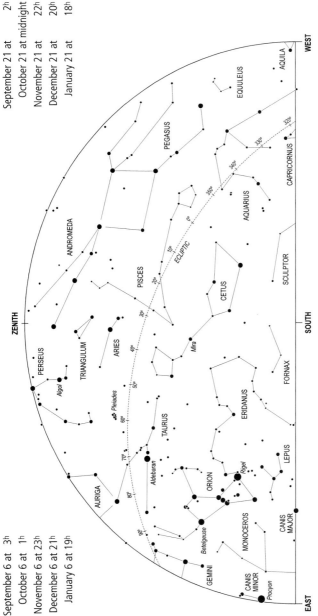

September 6 at    3ʰ
October 6 at    1ʰ
November 6 at  23ʰ
December 6 at  21ʰ
January 6 at  19ʰ

WEST

AQUILA

EQUULEUS

PEGASUS

ANDROMEDA

320°

330°

340°

CAPRICORNUS

350°

AQUARIUS

0°

ECLIPTIC

10°

PISCES

SCULPTOR

20°

CETUS

ZENITH

30°

SOUTH

PERSEUS

Algol

TRIANGULUM

ARIES

40°

Mira

FORNAX

Pleiades

50°

60°

ERIDANUS

TAURUS

70°

Aldebaran

LEPUS

AURIGA

80°

ORION

Rigel

Betelgeuse

MONOCEROS

CANIS
MAJOR

90°

GEMINI

CANIS
MINOR

Procyon

EAST

# 12N

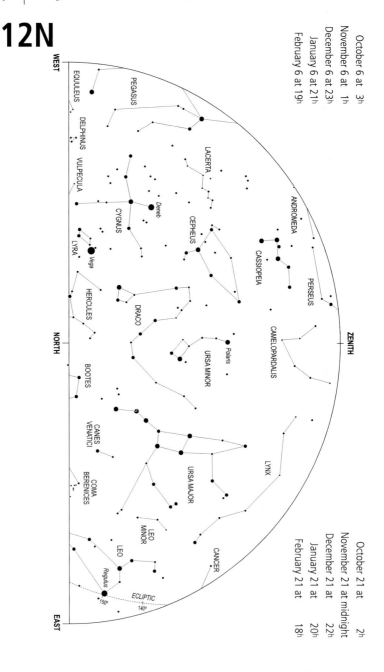

October 6 at 3h
November 6 at 1h
December 6 at 23h
January 6 at 21h
February 6 at 19h

October 21 at 2h
November 21 at midnight
December 21 at 22h
January 21 at 20h
February 21 at 18h

# 12S

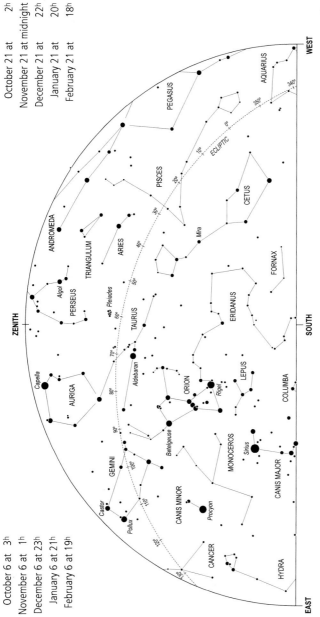

October 21 at 2ʰ
November 21 at midnight
December 21 at 22ʰ
January 21 at 20ʰ
February 21 at 18ʰ

October 6 at 3ʰ
November 6 at 1ʰ
December 6 at 23ʰ
January 6 at 21ʰ
February 6 at 19ʰ

WEST

ZENITH

SOUTH

EAST

AQUARIUS
PEGASUS
PISCES
CETUS
ANDROMEDA
TRIANGULUM
ARIES
FORNAX
PERSEUS
*Algol*
ERIDANUS
*Mira*
*Pleiades*
TAURUS
*Aldebaran*
*Capella*
AURIGA
ORION
LEPUS
*Rigel*
COLUMBA
*Betelgeuse*
GEMINI
MONOCEROS
*Sirius*
*Castor*
*Pollux*
CANIS MINOR
CANIS MAJOR
*Procyon*
CANCER
HYDRA

ECLIPTIC

# Southern Star Charts

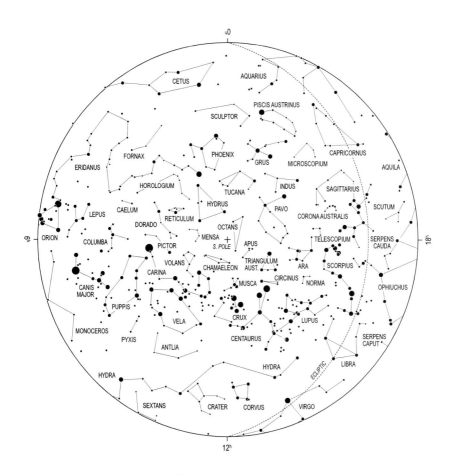

**Southern Hemisphere**

Note that the markers at 0ʰ, 6ʰ, 12ʰ and 18ʰ indicate hours of Right Ascension.

# 1N

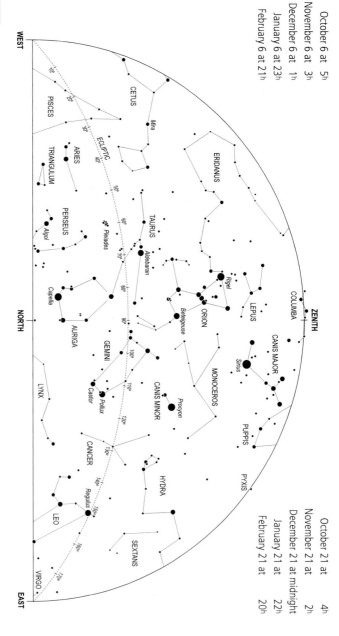

WEST

CETUS

PISCES

Mira

ECLIPTIC

ERIDANUS

TRIANGULUM

ARIES

10°

20°

30°

40°

50°

60°

70°

80°

90°

100°

110°

120°

130°

140°

150°

160°

170°

Pleiades

PERSEUS

Algol

TAURUS

Aldebaran

Capella

AURIGA

GEMINI

Castor

Pollux

LYNX

Rigel

ORION

Betelgeuse

LEPUS

COLUMBA

ZENITH

CANIS MAJOR

Sirius

MONOCEROS

CANIS MINOR

Procyon

PUPPIS

PYXIS

NORTH

CANCER

Regulus

LEO

HYDRA

SEXTANS

VIRGO

EAST

**1S**

October 21 at 4ʰ
November 21 at 2ʰ
December 21 at midnight
January 21 at 22ʰ
February 21 at 20ʰ

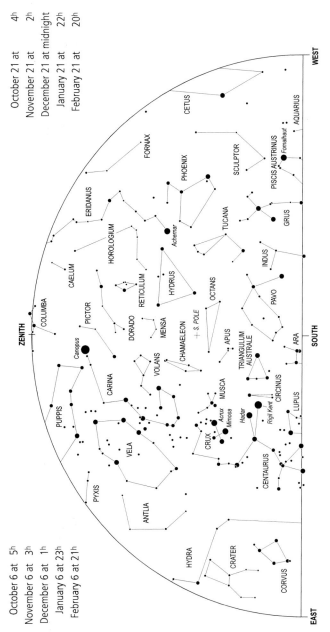

WEST

CETUS

AQUARIUS

FORNAX

PHOENIX

SCULPTOR

PISCIS AUSTRINUS
*Fomalhaut*

ERIDANUS

TUCANA

GRUS

HOROLOGIUM

*Achernar*

INDUS

CAELUM

RETICULUM

HYDRUS

OCTANS

PAVO

COLUMBA

PICTOR

DORADO

MENSA

+ S. POLE

ZENITH

*Canopus*

VOLANS

CHAMAELEON

APUS

TRIANGULUM
AUSTRALE

ARA

SOUTH

CARINA

MUSCA

PUPPIS

*Actrux*
*Mimosa*

CRUX

CIRCINUS

*Hadar*

*Rigil Kent*

LUPUS

VELA

CENTAURUS

PYXIS

ANTLIA

HYDRA

CRATER

CORVUS

EAST

October 6 at 5ʰ
November 6 at 3ʰ
December 6 at 1ʰ
January 6 at 23ʰ
February 6 at 21ʰ

## 2N

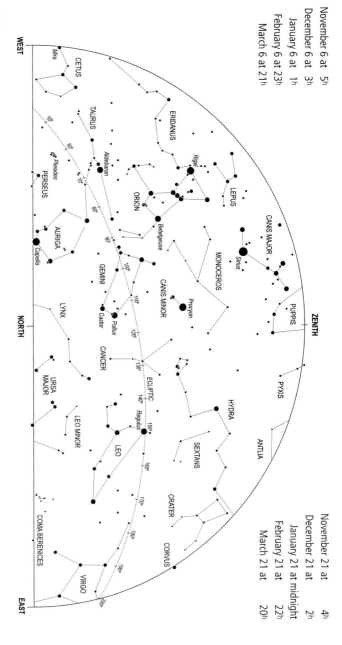

November 6 at 5h
December 6 at 3h
January 6 at 1h
February 6 at 23h
March 6 at 21h

November 21 at 4h
December 21 at 2h
January 21 at midnight
February 21 at 22h
March 21 at 20h

November 21 at 4h
December 21 at 2h
January 21 at midnight
February 21 at 22h
March 21 at 20h

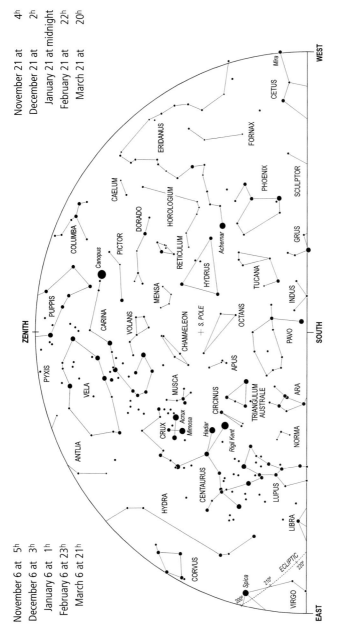

WEST

Mira

CETUS

ERIDANUS

FORNAX

PHOENIX

SCULPTOR

CAELUM

DORADO

HOROLOGIUM

COLUMBA

PICTOR

RETICULUM

Achernar

GRUS

Canopus

MENSA

HYDRUS

TUCANA

INDUS

PUPPIS

CARINA

VOLANS

CHAMAELEON

+ S. POLE

OCTANS

PAVO

SOUTH

ZENITH

PYXIS

APUS

VELA

MUSCA

CIRCINUS

TRIANGULUM
AUSTRALE

ARA

ANTLIA

CRUX

Acrux

Mimosa

Hadar

Rigil Kent

NORMA

CENTAURUS

LUPUS

HYDRA

LIBRA

CORVUS

VIRGO

Spica

ECLIPTIC

210°

220°

200°

EAST

November 6 at 5h
December 6 at 3h
January 6 at 1h
February 6 at 23h
March 6 at 21h

# 3N

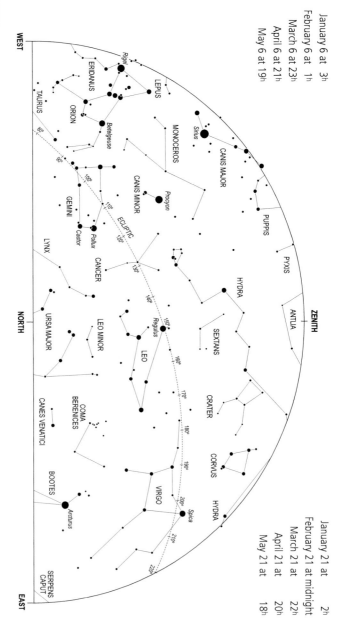

January 6 at 3h
February 6 at 1h
March 6 at 23h
April 6 at 21h
May 6 at 19h

January 21 at 2h
February 21 at midnight
March 21 at 22h
April 21 at 20h
May 21 at 18h

**3S**

January 21 at 2ʰ
February 21 at midnight
March 21 at 22ʰ
April 21 at 20ʰ
May 21 at 18ʰ

January 6 at 3ʰ
February 6 at 1ʰ
March 6 at 23ʰ
April 6 at 21ʰ
May 6 at 19ʰ

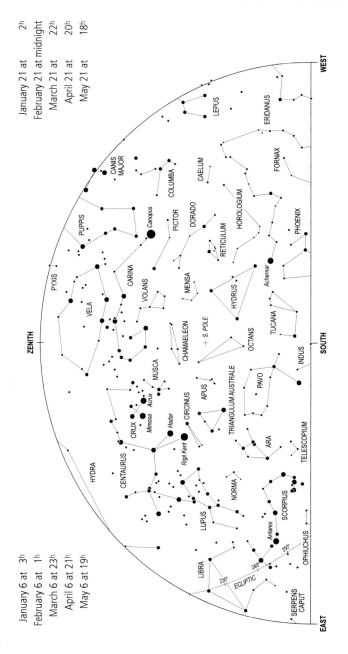

WEST

ZENITH

SOUTH

EAST

LEPUS
ERIDANUS
CANIS MAJOR
COLUMBA
CAELUM
FORNAX
PUPPIS
Canopus
PICTOR
DORADO
HOROLOGIUM
PHOENIX
PYXIS
CARINA
RETICULUM
Achernar
VELA
VOLANS
MENSA
HYDRUS
TUCANA
+ S. POLE
CHAMAELEON
OCTANS
INDUS
MUSCA
Acrux
APUS
PAVO
CRUX
Mimosa
Hadar
CIRCINUS
TRIANGULUM AUSTRALE
Rigil Kent
ARA
CENTAURUS
TELESCOPIUM
HYDRA
NORMA
SCORPIUS
LUPUS
Antares
OPHIUCHUS
LIBRA
250°
240°
230°
ECLIPTIC
SERPENS CAPUT

# 4N

February 6 at 3h
March 6 at 1h
April 6 at 23h
May 6 at 21h
June 6 at 19h

February 21 at 2h
March 21 at midnight
April 21 at 22h
May 21 at 20h
June 21 at 18h

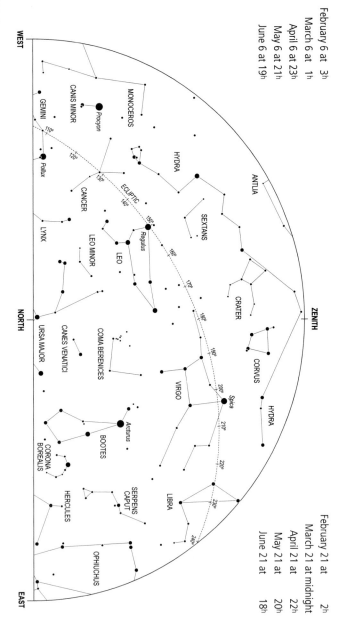

WEST

GEMINI
CANIS MINOR
Procyon
MONOCEROS
Pollux
110°
120°
CANCER
130°
ECLIPTIC
140°
LYNX
LEO MINOR
150°
Regulus
LEO
HYDRA
SEXTANS
160°
ANTLIA
CRATER
170°
CORVUS
180°
HYDRA
NORTH
URSA MAJOR
CANES VENATICI
COMA BERENICES
190°
VIRGO
200°
Spica
ZENITH
Arcturus
210°
BOOTES
CORONA BOREALIS
220°
HERCULES
SERPENS CAPUT
LIBRA
230°
OPHIUCHUS
240°
EAST

**4S**

February 21 at 2ʰ
March 21 at midnight
April 21 at 22ʰ
May 21 at 20ʰ
June 21 at 18ʰ

February 6 at 3ʰ
March 6 at 1ʰ
April 6 at 23ʰ
May 6 at 21ʰ
June 6 at 19ʰ

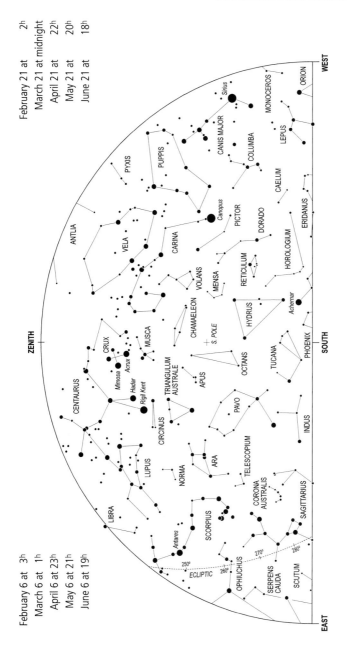

WEST

ORION

MONOCEROS

LEPUS

CANIS MAJOR

Sirius

COLUMBA

CAELUM

PUPPIS

PYXIS

PICTOR

Canopus

DORADO

ERIDANUS

ANTLIA

VELA

CARINA

HOROLOGIUM

VOLANS

RETICULUM

MENSA

Achernar

CHAMAELEON

HYDRUS

ZENITH

MUSCA

CRUX

S. POLE

PHOENIX

SOUTH

Acrux

Mimosa

Hadar

Rigil Kent

CENTAURUS

TRIANGULUM
AUSTRALE

APUS

OCTANS

TUCANA

CIRCINUS

PAVO

INDUS

LUPUS

NORMA

ARA

TELESCOPIUM

LIBRA

SCORPIUS

CORONA
AUSTRALIS

SAGITTARIUS

Antares

270°

280°

250°

260°

ECLIPTIC

OPHIUCHUS

SERPENS
CAUDA

SCUTUM

EAST

# 5N

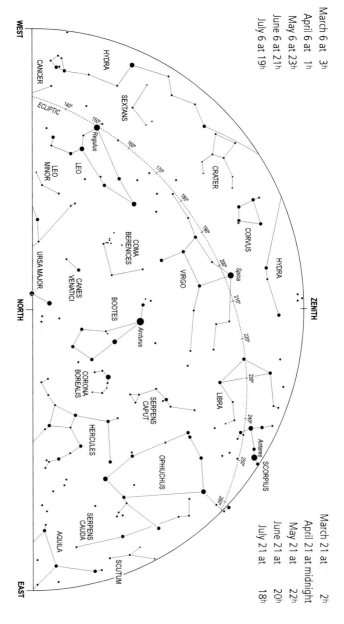

March 6 at 3h
April 6 at 1h
May 6 at 23h
June 6 at 21h
July 6 at 19h

March 21 at 2h
April 21 at midnight
May 21 at 22h
June 21 at 20h
July 21 at 18h

**5S**

March 21 at 2ʰ
April 21 at midnight
May 21 at 22ʰ
June 21 at 20ʰ
July 21 at 18ʰ

March 6 at 3ʰ
April 6 at 1ʰ
May 6 at 23ʰ
June 6 at 21ʰ
July 6 at 19ʰ

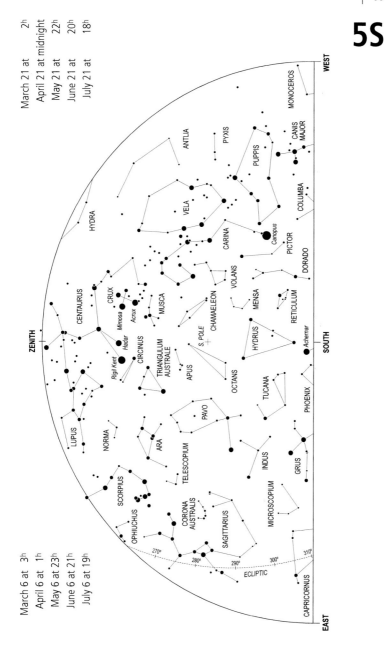

WEST

ZENITH

SOUTH

EAST

MONOCEROS
ANTLIA
PYXIS
PUPPIS
CANIS MAJOR
VELA
COLUMBA
HYDRA
CARINA
Canopus
PICTOR
DORADO
CENTAURUS
CRUX
Mimosa
Acrux
MUSCA
VOLANS
CHAMAELEON
MENSA
RETICULUM
Achernar
Hadar
Rigil Kent
CIRCINUS
TRIANGULUM AUSTRALE
S. POLE
HYDRUS
APUS
OCTANS
LUPUS
NORMA
ARA
PAVO
TUCANA
PHOENIX
TELESCOPIUM
INDUS
GRUS
SCORPIUS
OPHIUCHUS
CORONA AUSTRALIS
SAGITTARIUS
MICROSCOPIUM
270°
280°
290°
300°
310°
ECLIPTIC
CAPRICORNUS

# 6N

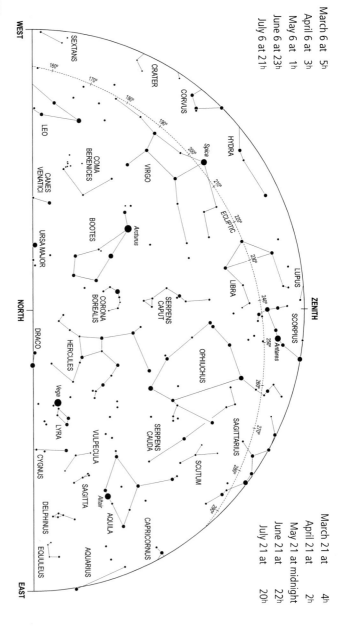

March 6 at 5h
April 6 at 3h
May 6 at 1h
June 6 at 23h
July 6 at 21h

WEST

SEXTANS
CRATER
CORVUS
HYDRA
LEO
COMA BERENICES
VIRGO
Spica
CANES VENATICI
ECLIPTIC
URSA MAJOR
BOOTES
Arcturus
LIBRA
LUPUS
ZENITH
SCORPIUS
Antares
CORONA BOREALIS
SERPENS CAPUT
DRACO
HERCULES
OPHIUCHUS
NORTH
Vega
LYRA
SERPENS CAUDA
SAGITTARIUS
VULPECULA
CYGNUS
SCUTUM
SAGITTA
Altair
AQUILA
CAPRICORNUS
DELPHINUS
AQUARIUS
EQUULEUS
EAST

160° 170° 180° 190° 200° 210° 220° 230° 240° 250° 260° 270° 280° 290° 300°

March 21 at 4h
April 21 at 2h
May 21 at midnight
June 21 at 22h
July 21 at 20h

March 21 at 4ʰ
April 21 at 2ʰ
May 21 at midnight
June 21 at 22ʰ
July 21 at 20ʰ

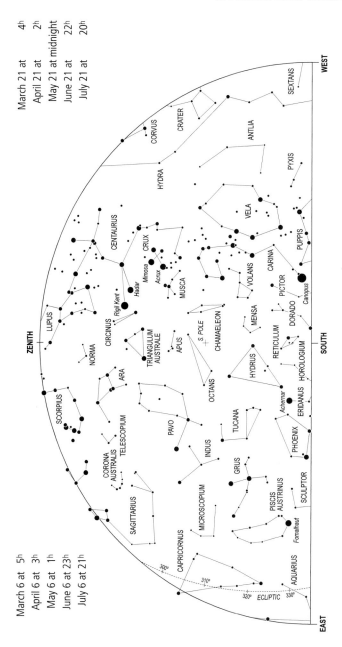

WEST

SEXTANS

CORVUS

CRATER

ANTLIA

HYDRA

PYXIS

CENTAURUS

VELA

PUPPIS

CRUX

Mimosa

Acrux

CARINA

MUSCA

VOLANS

PICTOR

Hadar

DORADO

Canopus

Rigil Kent

LUPUS

CIRCINUS

S. POLE

CHAMAELEON

MENSA

RETICULUM

NORMA

TRIANGULUM
AUSTRALE

APUS

ZENITH

ARA

OCTANS

HYDRUS

HOROLOGIUM

SCORPIUS

PAVO

TUCANA

Achernar

ERIDANUS

SOUTH

CORONA
AUSTRALIS

TELESCOPIUM

INDUS

PHOENIX

SAGITTARIUS

MICROSCOPIUM

GRUS

SCULPTOR

PISCIS
AUSTRINUS

CAPRICORNUS

Fomalhaut

AQUARIUS

300°

310°

320° ECLIPTIC 330°

EAST

March 6 at 5ʰ
April 6 at 3ʰ
May 6 at 1ʰ
June 6 at 23ʰ
July 6 at 21ʰ

# 7N

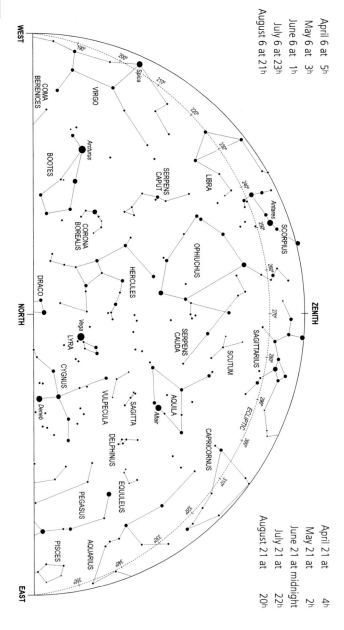

April 6 at 5h
May 6 at 3h
June 6 at 1h
July 6 at 23h
August 6 at 21h

April 21 at 4h
May 21 at 2h
June 21 at midnight
July 21 at 22h
August 21 at 20h

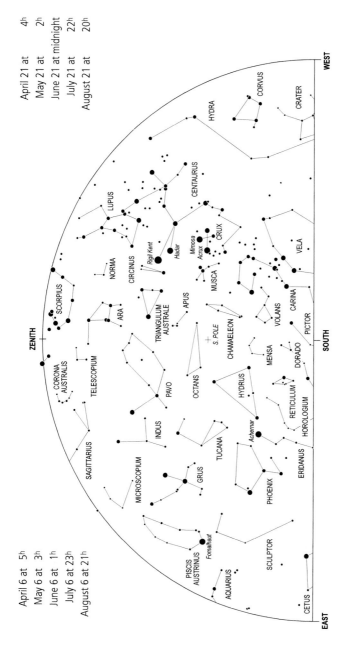

April 21 at 4h
May 21 at 2h
June 21 at midnight
July 21 at 22h
August 21 at 20h

WEST

CORVUS
CRATER
HYDRA
CENTAURUS
LUPUS
NORMA
CIRCINUS
Rigil Kent
Hadar
Mimosa
Acrux
CRUX
MUSCA
VELA
APUS
ARA
TRIANGULUM AUSTRALE
VOLANS
CARINA
PICTOR
SCORPIUS
ZENITH
S. POLE
CHAMAELEON
MENSA
DORADO
SOUTH
CORONA AUSTRALIS
TELESCOPIUM
PAVO
OCTANS
HYDRUS
RETICULUM
HOROLOGIUM
SAGITTARIUS
INDUS
Achernar
TUCANA
ERIDANUS
MICROSCOPIUM
GRUS
PHOENIX
Fomalhaut
SCULPTOR
PISCIS AUSTRINUS
AQUARIUS
CETUS
EAST

April 6 at 5h
May 6 at 3h
June 6 at 1h
July 6 at 23h
August 6 at 21h

**7S**

# 8N

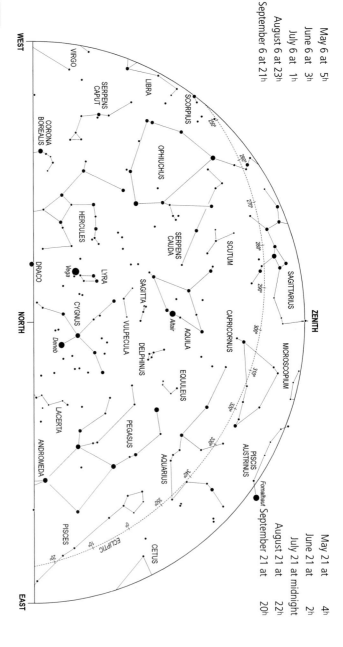

**8S**

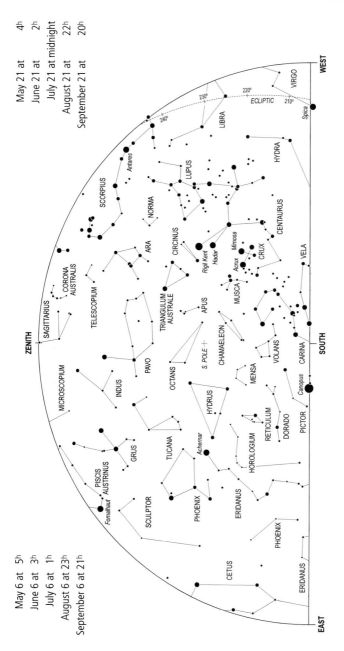

May 21 at   4ʰ
June 21 at   2ʰ
July 21 at midnight
August 21 at   22ʰ
September 21 at   20ʰ

May 6 at   5ʰ
June 6 at   3ʰ
July 6 at   1ʰ
August 6 at   23ʰ
September 6 at   21ʰ

# 9N

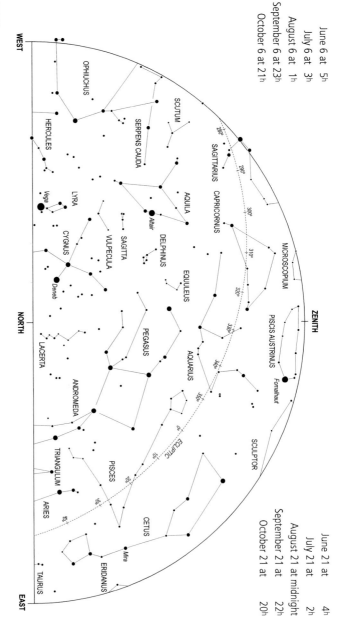

June 6 at 5h
July 6 at 3h
August 6 at 1h
September 6 at 23h
October 6 at 21h

June 21 at 4h
July 21 at 2h
August 21 at midnight
September 21 at 22h
October 21 at 20h

WEST

NORTH

EAST

ZENITH

OPHIUCHUS
SCUTUM
SERPENS CAUDA
HERCULES
SAGITTARIUS
Vega
LYRA
AQUILA
CAPRICORNUS
Altair
280°
290°
300°
310°
MICROSCOPIUM
CYGNUS
SAGITTA
DELPHINUS
VULPECULA
320°
PISCIS AUSTRINUS
Deneb
EQQULEUS
330°
340°
Fomalhaut
LACERTA
PEGASUS
AQUARIUS
350°
360°
SCULPTOR
ANDROMEDA
ECLIPTIC
0°
TRIANGULUM
PISCES
10°
20°
ARIES
30°
CETUS
40°
Mira
TAURUS
ERIDANUS

**9S**

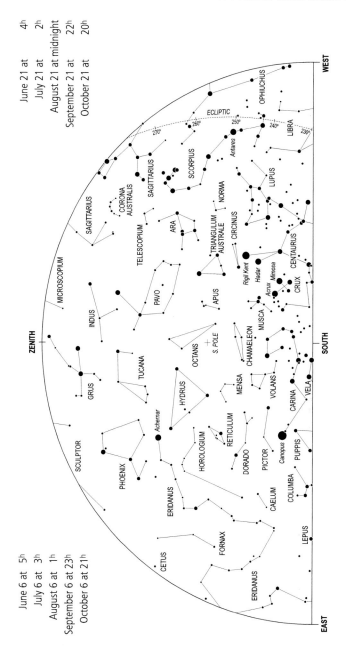

June 21 at 4ʰ
July 21 at 2ʰ
August 21 at midnight
September 21 at 22ʰ
October 21 at 20ʰ

June 6 at 5ʰ
July 6 at 3ʰ
August 6 at 1ʰ
September 6 at 23ʰ
October 6 at 21ʰ

WEST

ZENITH

SOUTH

EAST

ECLIPTIC

270° 260° 250° 240° 230°

OPHIUCHUS
LIBRA
Antares
SCORPIUS
LUPUS
SAGITTARIUS
CORONA AUSTRALIS
NORMA
TELESCOPIUM
TRIANGULUM AUSTRALE
CIRCINUS
SAGITTARIUS
ARA
CENTAURUS
MICROSCOPIUM
PAVO
APUS
Rigil Kent
Hadar
Mimosa
Acrux
CRUX
MUSCA
INDUS
S. POLE
OCTANS
CHAMAELEON
GRUS
TUCANA
HYDRUS
MENSA
VOLANS
CARINA
VELA
SCULPTOR
Achernar
RETICULUM
DORADO
PICTOR
Canopus
PUPPIS
PHOENIX
HOROLOGIUM
ERIDANUS
CAELUM
COLUMBA
CETUS
FORNAX
LEPUS
ERIDANUS

# 10N

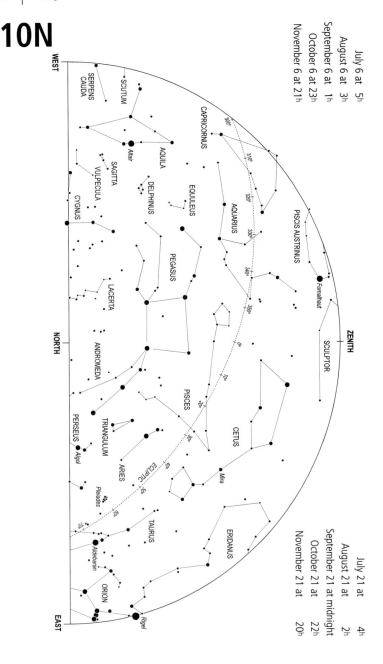

WEST

SERPENS CAUDA
SCUTUM
CAPRICORNUS
300°
AQUILA
Altair
310°
SAGITTA
VULPECULA
DELPHINUS
EQUULEUS
320°
AQUARIUS
PISCIS AUSTRINUS
CYGNUS
330°
PEGASUS
340°
Fomalhaut
LACERTA
350°
SCULPTOR
ZENITH
ANDROMEDA
0°
10°
PISCES
CETUS
20°
TRIANGULUM
PERSEUS
30°
Algol
ARIES
Mira
40°
NORTH
Pleiades
ECLIPTIC
50°
60°
TAURUS
ERIDANUS
70°
Aldebaran
ORION
Rigel
EAST

# 10S

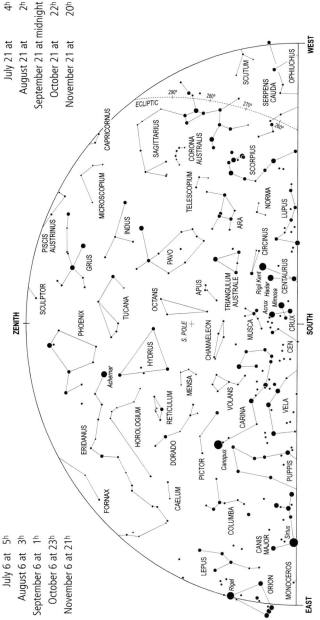

July 21 at 4ʰ
August 21 at 2ʰ
September 21 at midnight
October 21 at 22ʰ
November 21 at 20ʰ

July 6 at 5ʰ
August 6 at 3ʰ
September 6 at 1ʰ
October 6 at 23ʰ
November 6 at 21ʰ

WEST

EAST

SOUTH

ZENITH

OPHIUCHUS
SCUTUM
SERPENS CAUDA
SAGITTARIUS
CORONA AUSTRALIS
SCORPIUS
NORMA
LUPUS
ARA
TELESCOPIUM
CIRCINUS
CENTAURUS
Rigel Kent
Hadar
Mimosa
Acrux
CRUX
CEN
MUSCA
TRIANGULUM AUSTRALE
APUS
PAVO
INDUS
MICROSCOPIUM
CAPRICORNUS
PISCIS AUSTRINUS
GRUS
TUCANA
OCTANS
CHAMAELEON
S. POLE
HYDRUS
MENSA
VOLANS
CARINA
VELA
PUPPIS
SCULPTOR
PHOENIX
Achernar
ERIDANUS
RETICULUM
HOROLOGIUM
DORADO
PICTOR
Canopus
CAELUM
FORNAX
COLUMBA
CANIS MAJOR
Sirius
LEPUS
Rigel
ORION
MONOCEROS

ECLIPTIC
290°
280°
270°
280°

# 11N

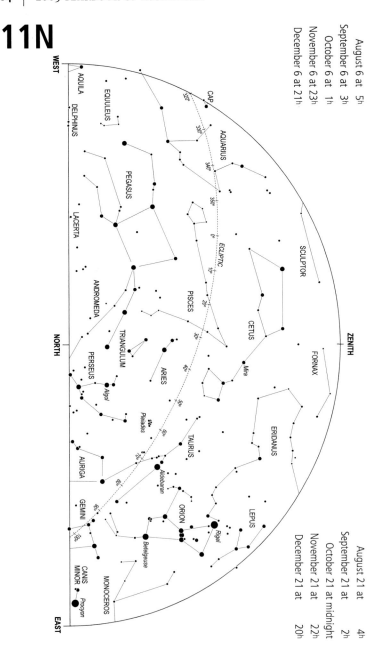

August 6 at 5h
September 6 at 3h
October 6 at 1h
November 6 at 23h
December 6 at 21h

August 21 at 4h
September 21 at 2h
October 21 at midnight
November 21 at 22h
December 21 at 20h

# 11S

August 21 at 4ʰ
September 21 at 2ʰ
October 21 at midnight
November 21 at 22ʰ
December 21 at 20ʰ

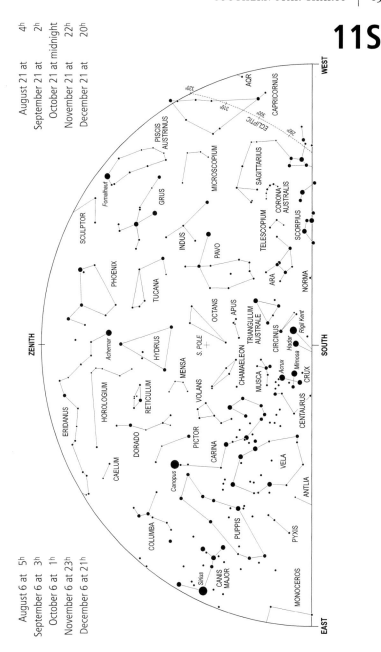

WEST

AQR
CAPRICORNUS
320°
PISCIS
AUSTRINUS
310°
ECLIPTIC
300°
290°
SAGITTARIUS
MICROSCOPIUM
Fomalhaut
GRUS
CORONA
AUSTRALIS
SCULPTOR
TELESCOPIUM
SCORPIUS
INDUS
PAVO
PHOENIX
ARA
NORMA
TUCANA
ZENITH
OCTANS
APUS
TRIANGULUM
AUSTRALE
Rigil Kent
CIRCINUS
Achernar
HYDRUS
S. POLE
CHAMAELEON
Hadar
Acrux
Mimosa
CRUX
ERIDANUS
MENSA
MUSCA
SOUTH
HOROLOGIUM
VOLANS
CENTAURUS
RETICULUM
CAELUM
DORADO
PICTOR
CARINA
VELA
ANTLIA
Canopus
COLUMBA
PUPPIS
PYXIS
Sirius
CANIS
MAJOR
MONOCEROS

EAST

August 6 at 5ʰ
September 6 at 3ʰ
October 6 at 1ʰ
November 6 at 23ʰ
December 6 at 21ʰ

# 12N

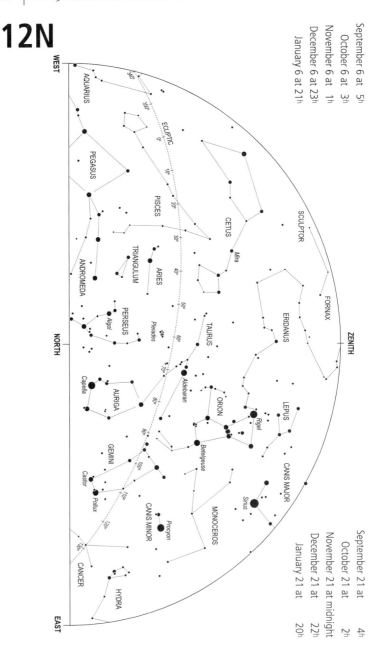

September 6 at 5h
October 6 at 3h
November 6 at 1h
December 6 at 23h
January 6 at 21h

September 21 at 4h
October 21 at 2h
November 21 at midnight
December 21 at 22h
January 21 at 20h

# 12S

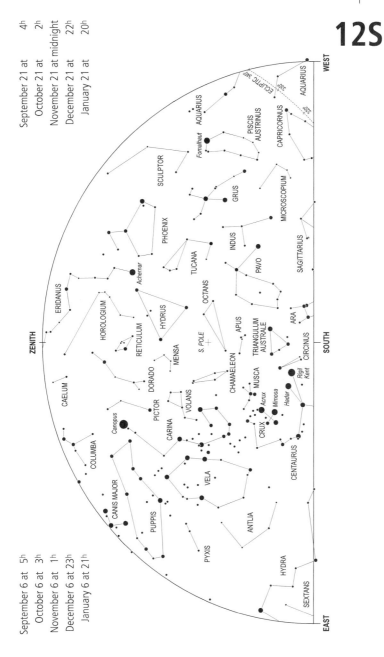

WEST

ZENITH

SOUTH

EAST

ECLIPTIC 34°

35°

35°

AQUARIUS

AQUARIUS

PISCIS AUSTRINUS

CAPRICORNUS

Fomalhaut

SCULPTOR

GRUS

MICROSCOPIUM

PHOENIX

INDUS

SAGITTARIUS

TUCANA

PAVO

Achernar

OCTANS

ERIDANUS

HYDRUS

ARA

HOROLOGIUM

RETICULUM

S. POLE

APUS

TRIANGULUM AUSTRALE

MENSA

CHAMAELEON

CIRCINUS

CAELUM

DORADO

MUSCA

Rigil Kent

PICTOR

VOLANS

Acrux

Mimosa

Hadar

COLUMBA

Canopus

CARINA

CRUX

CENTAURUS

CANIS MAJOR

VELA

PUPPIS

ANTLIA

PYXIS

HYDRA

SEXTANS

# The Planets and the Ecliptic

The paths of the planets about the Sun all lie close to the plane of the ecliptic, which is marked for us in the sky by the apparent path of the Sun among the stars, and is shown on the star charts by a broken line. The Moon and naked-eye planets will always be found close to this line, never departing from it by more than about 7°. Thus the planets are most favourably placed for observation when the ecliptic is well displayed, and this means that it should be as high in the sky as possible. This avoids the difficulty of finding a clear horizon, and also overcomes the problem of atmospheric absorption, which greatly reduces the light of the stars. Thus a star at an altitude of 10° suffers a loss of 60 per cent of its light, which corresponds to a whole magnitude; at an altitude of only 4°, the loss may amount to two magnitudes.

The position of the ecliptic in the sky is therefore of great importance, and since it is tilted at about 23.5° to the equator, it is only at certain times of the day or year that it is displayed to the best advantage. It will be realized that the Sun (and therefore the ecliptic) is at its highest in the sky at noon in midsummer, and at its lowest at noon in midwinter. Allowing for the daily motion of the sky, it follows that the ecliptic is highest at midnight in winter, at sunset in the spring, at noon in summer and at sunrise in the autumn. Hence, these are the best times to see the planets. Thus, if Venus is an evening object in the western sky after sunset, it will be seen to best advantage if this occurs in the spring, when the ecliptic is high in the sky and slopes down steeply to the horizon. This means that the planet is not only higher in the sky, but will remain for a much longer period above the horizon. For similar reasons, a morning object will be seen at its best on autumn mornings before sunrise, when the ecliptic is high in the east. The outer planets, which can come to opposition (i.e. opposite the Sun), are best seen when opposition occurs in the winter months, when the ecliptic is high in the sky at midnight.

The seasons are reversed in the Southern Hemisphere, spring beginning at the September Equinox, when the Sun crosses the equator on its way south, summer beginning at the December Solstice, when the

Sun is highest in the southern sky, and so on. Thus, the times when the ecliptic is highest in the sky, and therefore best placed for observing the planets, may be summarized as follows:

|  | Midnight | Sunrise | Noon | Sunset |
|---|---|---|---|---|
| Northern latitudes | December | September | June | March |
| Southern latitudes | June | March | December | September |

In addition to the apparent daily rotation of the celestial sphere from east to west, the planets have a motion of their own among the stars. The apparent movement is generally *direct*, i.e. to the east, in the direction of increasing longitude, but for a certain period (which depends on the distance of the planet) this apparent motion is reversed. With the outer planets this *retrograde* motion occurs about the time of opposition. Owing to the different inclination of the orbits of these planets, the actual effect is to cause the apparent path to form a loop, or sometimes an S-shaped curve. The same effect is present in the motion of the inferior planets, Mercury and Venus, but it is not so obvious, since it always occurs at the time of inferior conjunction.

The *inferior planets*, Mercury and Venus, move in smaller orbits than that of the Earth, and so are always seen near the Sun. They are most obvious at the times of greatest angular distance from the Sun (greatest elongation), which may reach 28° for Mercury, and 47° for Venus. They are seen as evening objects in the western sky after sunset (at eastern elongations) or as morning objects in the eastern sky before sunrise (at western elongations). The succession of phenomena, conjunctions and elongations always follows the same order, but the intervals between them are not equal. Thus, if either planet is moving round the far side of its orbit its motion will be to the east, in the same direction in which the Sun appears to be moving. It therefore takes much longer for the planet to overtake the Sun – that is, to come to superior conjunction – than it does when moving round to inferior conjunction, between Sun and Earth. The intervals given in the table at the top of p. 70 are average values; they remain fairly constant in the case of Venus, which travels in an almost circular orbit. In the case of Mercury, however, conditions vary widely because of the great eccentricity and inclination of the planet's orbit.

|  | | Mercury | Venus |
|---|---|---|---|
| Inferior Conjunction | to Elongation West | 22 days | 72 days |
| Elongation West | to Superior Conjunction | 36 days | 220 days |
| Superior Conjunction | to Elongation East | 35 days | 220 days |
| Elongation East | to Inferior Conjunction | 22 days | 72 days |

The greatest brilliancy of Venus always occurs about thirty-six days before or after inferior conjunction. This will be about a month after greatest eastern elongation (as an evening object), or a month before greatest western elongation (as a morning object). No such rule can be given for Mercury, because its distances from the Earth and the Sun can vary over a wide range.

*Mercury* is not likely to be seen unless a clear horizon is available. It is seldom as much as 10° above the horizon in the twilight sky in northern temperate latitudes, but this figure is often exceeded in the Southern Hemisphere. This favourable condition arises because the maximum elongation of 28° can occur only when the planet is at aphelion (farthest from the Sun), and it then lies well south of the equator. Northern observers must be content with smaller elongations, which may be as little as 18° at perihelion. In general, it may be said that the most favourable times for seeing Mercury as an evening object will be in spring, some days before greatest eastern elongation; in autumn, it may be seen as a morning object some days after greatest western elongation.

*Venus* is the brightest of the planets and may be seen on occasions in broad daylight. Like Mercury, it is alternately a morning and an evening object, and it will be highest in the sky when it is a morning object in autumn, or an evening object in spring. Venus is to be seen at its best as an evening object in northern latitudes when eastern elongation occurs in June. The planet is then well north of the Sun in the preceding spring months, and is a brilliant object in the evening sky over a long period. In the Southern Hemisphere a November elongation is best. For similar reasons, Venus gives a prolonged display as a morning object in the months following western elongation in October (in northern latitudes) or in June (in the Southern Hemisphere).

The *superior planets*, which travel in orbits larger than that of the Earth, differ from Mercury and Venus in that they can be seen opposite the Sun in the sky. The superior planets are morning objects after conjunction with the Sun, rising earlier each day until they come to opposition. They will then be nearest to the Earth (and therefore at their brightest), and will be on the meridian at midnight, due south in northern latitudes, but due north in the Southern Hemisphere. After opposition they are evening objects, setting earlier each evening until they set in the west with the Sun at the next conjunction. The difference in brightness from one opposition to another is most noticeable in the case of Mars, whose distance from Earth can vary considerably and rapidly. The other superior planets are at such great distances that there is very little change in brightness from one opposition to the next. The effect of altitude is, however, of some importance, for at a December opposition in northern latitudes the planets will be among the stars of Taurus or Gemini, and can then be at an altitude of more than 60° in southern England. At a summer opposition, when the planet is in Sagittarius, it may only rise to about 15° above the southern horizon, and so makes a less impressive appearance. In the Southern Hemisphere the reverse conditions apply, a June opposition being the best, with the planet in Sagittarius at an altitude which can reach 80° above the northern horizon for observers in South Africa.

*Mars*, whose orbit is appreciably eccentric, comes nearest to the Earth at oppositions at the end of August, as it was in 2003. It may then be brighter even than Jupiter, but rather low in the sky in Aquarius for northern observers, though very well placed for those in southern latitudes. These favourable oppositions occur every fifteen or seventeen years (e.g. in 1988, 2003 and 2018). In the Northern Hemisphere the planet is probably better seen at oppositions in the autumn or winter months, when it is higher in the sky – such as in 2005, when opposition is in early November. Oppositions of Mars occur at an average interval of 780 days, and during this time the planet makes a complete circuit of the sky.

*Jupiter* is always a bright planet, and comes to opposition a month later each year, having moved, roughly speaking, from one Zodiacal constellation to the next.

*Saturn* moves much more slowly than Jupiter, and may remain in the same constellation for several years. The brightness of Saturn depends on the aspects of its rings, as well as on the distance from Earth and Sun. The Earth passed through the plane of Saturn's rings in 1995 and 1996, when they appeared edge-on; we saw them at maximum opening, and Saturn at its brightest, in 2002. The rings will next appear edge-on in 2009.

*Uranus* and *Neptune* are both visible with binoculars or a small telescope, but you will need a finder chart to help locate them, while *Pluto* is hardly likely to attract the attention of observers without adequate telescopes.

# Phases of the Moon in 2005

| New Moon | | | | First Quarter | | | | Full Moon | | | | Last Quarter | | | |
|---|---|---|---|---|---|---|---|---|---|---|---|---|---|---|---|
| | d | h | m | | d | h | m | | d | h | m | | d | h | m |
| | | | | | | | | | | | | Jan | 3 | 17 | 46 |
| Jan | 10 | 12 | 03 | Jan | 17 | 06 | 57 | Jan | 25 | 10 | 32 | Feb | 2 | 07 | 27 |
| Feb | 8 | 22 | 28 | Feb | 16 | 00 | 16 | Feb | 24 | 04 | 54 | Mar | 3 | 17 | 36 |
| Mar | 10 | 09 | 10 | Mar | 17 | 19 | 19 | Mar | 25 | 20 | 58 | Apr | 2 | 00 | 50 |
| Apr | 8 | 20 | 32 | Apr | 16 | 14 | 37 | Apr | 24 | 10 | 06 | May | 1 | 06 | 24 |
| May | 8 | 08 | 45 | May | 16 | 08 | 57 | May | 23 | 20 | 18 | May | 30 | 11 | 47 |
| June | 6 | 21 | 55 | June | 15 | 01 | 22 | June | 22 | 04 | 14 | June | 28 | 18 | 23 |
| July | 6 | 12 | 02 | July | 14 | 15 | 20 | July | 21 | 11 | 00 | July | 28 | 03 | 19 |
| Aug | 5 | 03 | 05 | Aug | 13 | 02 | 39 | Aug | 19 | 17 | 53 | Aug | 26 | 15 | 18 |
| Sept | 3 | 18 | 45 | Sept | 11 | 11 | 37 | Sept | 18 | 02 | 01 | Sept | 25 | 06 | 41 |
| Oct | 3 | 10 | 28 | Oct | 10 | 19 | 01 | Oct | 17 | 12 | 14 | Oct | 25 | 01 | 17 |
| Nov | 2 | 01 | 25 | Nov | 9 | 01 | 57 | Nov | 16 | 00 | 58 | Nov | 23 | 22 | 11 |
| Dec | 1 | 15 | 01 | Dec | 8 | 09 | 36 | Dec | 15 | 16 | 16 | Dec | 23 | 19 | 36 |
| Dec | 31 | 03 | 12 | | | | | | | | | | | | |

All times are GMT

# Longitudes of the Sun, Moon and Planets in 2005

| Date | | Sun ° | Moon ° | Venus ° | Mars ° | Jupiter ° | Saturn ° |
|------|---|------|------|-------|------|---------|--------|
| January | 6 | 286 | 224 | 265 | 248 | 198 | 115 |
| | 21 | 301 | 73 | 284 | 258 | 199 | 113 |
| February | 6 | 317 | 276 | 304 | 269 | 199 | 112 |
| | 21 | 332 | 117 | 323 | 280 | 198 | 111 |
| March | 6 | 346 | 285 | 339 | 289 | 197 | 111 |
| | 21 | 0 | 125 | 358 | 300 | 196 | 110 |
| April | 6 | 16 | 339 | 18 | 312 | 194 | 111 |
| | 21 | 31 | 170 | 36 | 323 | 192 | 111 |
| May | 6 | 46 | 17 | 55 | 334 | 190 | 112 |
| | 21 | 60 | 203 | 73 | 344 | 189 | 113 |
| June | 6 | 75 | 65 | 93 | 356 | 189 | 115 |
| | 21 | 90 | 253 | 111 | 6 | 189 | 117 |
| July | 6 | 104 | 98 | 129 | 16 | 190 | 119 |
| | 21 | 118 | 292 | 148 | 26 | 192 | 121 |
| August | 6 | 134 | 143 | 167 | 35 | 194 | 123 |
| | 21 | 148 | 346 | 185 | 43 | 197 | 125 |
| September | 6 | 164 | 188 | 203 | 49 | 199 | 126 |
| | 21 | 178 | 37 | 221 | 53 | 202 | 128 |
| October | 6 | 193 | 223 | 238 | 53 | 206 | 129 |
| | 21 | 208 | 72 | 254 | 51 | 209 | 130 |
| November | 6 | 224 | 273 | 271 | 45 | 212 | 131 |
| | 21 | 239 | 117 | 285 | 41 | 216 | 131 |
| December | 6 | 254 | 312 | 295 | 38 | 219 | 131 |
| | 21 | 269 | 149 | 301 | 39 | 221 | 131 |

Longitude of  *Uranus*   338°        *Moon*: Longitude of ascending node
             *Neptune*  316°            Jan 1: 28°          Dec 31: 9°

Mercury moves so quickly among the stars that it is not possible to indicate its position on the star charts at convenient intervals. The monthly notes must be consulted for the best times at which the planet may be seen.

The positions of the other planets are given in the table on p. 74. This gives the apparent longitudes on dates that correspond to those of the star charts, and the position of the planet may at once be found near the ecliptic at the given longitude.

## EXAMPLES

*In the Southern Hemisphere, a planet is seen just west of the meridian in the northern sky early in November. Identify it.*

The southern star chart 12N shows the northern sky for 6 November at 01h, and shows longitudes 340° to 130° along the ecliptic, with a longitude of around 60° on the meridian, with decreasing longitudes to the west. Reference to the table on page 74 gives the longitude of Mars as 45°. No other naked-eye planets have longitudes anywhere near, so the identification has been made.

The positions of the Sun and Moon can be plotted on the star maps in the same manner as for the planets. The average daily motion of the Sun is 1°, and of the Moon 13°. For the Moon an indication of its position relative to the ecliptic may be obtained from a consideration of its longitude relative to that of the ascending node. The latter changes only slowly during the year, as will be seen from the values given on p. 74. Let us denote by $d$ the difference in longitude between the Moon and its ascending node. Then if $d = 0°$, 180° or 360°, the Moon is on the ecliptic. If $d = 90°$ the Moon is 5° north of the ecliptic, and if $d = 270°$ the Moon is 5° south of the ecliptic.

On 6 November the Moon's longitude is given in the table on p. 74 as 273° and the longitude of the node is found by interpolation to be about 12°. Thus $d = 261°$, and the Moon is about 5° south of the ecliptic. Its position may be plotted on northern star charts 6S, 7S and 8S, and on southern star charts 6N, 7N and 8N.

# Some Events in 2005

Jan   2   *Earth* at Perihelion
       10   New Moon
       13   *Saturn* at Opposition in Gemini
       25   Full Moon

Feb   3   *Neptune* in Conjunction with Sun
       8   New Moon
       14   *Mercury* at Superior Conjunction
       24   Full Moon
       25   *Uranus* in Conjunction with Sun

Mar   10   New Moon
       12   *Mercury* at Greatest Eastern Elongation (18°)
       20   Equinox (Spring Equinox in Northern Hemisphere)
       25   Full Moon
       29   *Mercury* at Inferior Conjunction
       31   *Venus* at Superior Conjunction

Apr   3   *Jupiter* at Opposition in Virgo
       8   New Moon
       8   Annular/Total Solar Eclipse
       24   Penumbral Lunar Eclipse
       24   Full Moon
       26   *Mercury* at Greatest Western Elongation (27°)

May   8   New Moon
       23   Full Moon

Jun   3   *Mercury* at Superior Conjunction
       6   New Moon
       14   *Pluto* at Opposition in Serpens
       21   Solstice (Summer Solstice in Northern Hemisphere)
       22   Full Moon

| Jul | 5 | *Earth* at Aphelion |
| | 6 | New Moon |
| | 9 | *Mercury* at Greatest Eastern Elongation (26°) |
| | 21 | Full Moon |
| | 23 | *Saturn* in Conjunction with Sun |

| Aug | 5 | New Moon |
| | 5 | *Mercury* at Inferior Conjunction |
| | 8 | *Neptune* at Opposition in Capricornus |
| | 19 | Full Moon |
| | 24 | *Mercury* at Greatest Western Elongation (18°) |

| Sep | 1 | *Uranus* at Opposition in Aquarius |
| | 3 | New Moon |
| | 18 | Full Moon |
| | 18 | *Mercury* at Superior Conjunction |
| | 22 | Equinox (Autumnal Equinox in Northern Hemisphere) |

| Oct | 3 | New Moon |
| | 3 | Annular Solar Eclipse |
| | 17 | Partial Lunar Eclipse |
| | 17 | Full Moon |
| | 22 | *Jupiter* in Conjunction with Sun |
| | 31 | *Mars* at Closest Approach to Earth |

| Nov | 2 | New Moon |
| | 3 | *Mercury* at Greatest Eastern Elongation (24°) |
| | 3 | *Venus* at Greatest Eastern Elongation (47°) |
| | 7 | *Mars* at Opposition in Aries |
| | 16 | Full Moon |
| | 23 | *Mercury* at Inferior Conjunction |

| Dec | 1 | New Moon |
| | 12 | *Mercury* at Greatest Western Elongation (21°) |
| | 15 | Full Moon |
| | 21 | Solstice (Winter Solstice in Northern Hemisphere) |
| | 31 | New Moon |

# Monthly Notes 2005

# January

*New Moon:* 10 January          *Full Moon:* 25 January

EARTH is at perihelion (nearest to the Sun) on 2 January at a distance of 147 million kilometres (91 million miles).

MERCURY, having reached greatest western elongation at the end of last month, is visible low in the south-eastern sky before dawn at the beginning of the month. Observers in the latitudes of the British Isles will only be able to detect it for the first week of the month at best, while those in equatorial and southern latitudes should be able to view it for the first four weeks of January. During this period its magnitude brightens slowly from −0.3 to −0.6. For the first half of the month Venus and Mercury are within a degree of each other, until Mercury commences to pull away to the east, finishing about 4° east of Venus by the end of its period of visibility.

VENUS, magnitude −3.9, is a brilliant object in the east-south-eastern sky before sunrise. However, for observers in the latitudes of the British Isles it is unlikely to be seen during the last few days of the month because of its very low altitude as it moves closer to the Sun.

MARS, magnitude +1.5, is a morning object, visible low in the south-eastern quadrant of the sky for an hour so before the increasing brightness of the pre-dawn sky inhibits observation. The planet moves slowly eastwards from Scorpius into Ophiuchus early in the month, passing 5° north of Antares on 7 January. Both objects are slightly reddish in colour, with Antares being the brighter by half a magnitude. Observers as far north as the British Isles will find it to be a very difficult object to detect even under very clear conditions, since it never gets more than about 10° high before dawn. Figure 9, given with the notes for May, shows the path of Mars during the early part of the year.

JUPITER, magnitude −2.1, is a brilliant object in the morning sky, in the constellation of Virgo. Before the end of the month it is visible in the east-south-eastern sky before midnight.

SATURN is at opposition on 13 January, and thus remains visible throughout the hours of darkness. Its distance from the Earth is then 1,208 million kilometres (751 million miles). Saturn is moving retrograde in the constellation of Gemini, about 7° south of Pollux. Figure 1 shows the path of Saturn against the constellations of Gemini and Cancer during 2005.

*Mars and Antares.* Because of its very variable distance from the Earth and its relatively small size (approximately half the diameter of the Earth), the brightness of Mars is very changeable. At its best it can outshine any star, and all of the planets apart from Venus; this will be the case in November 2005 when it is at opposition. When at its faintest, however, Mars fades to the second magnitude, and looks very much like a star. Unwary observers have often mistaken it for a nova.

This January, Mars is still a long way away from us, and is not yet striking, though it already ranks with the first-magnitude stars of the sky. As has been noted above, this month it passes just a few degrees north of the star Alpha Scorpii, also known as Antares – and the very name Antares means 'the Rival of Mars', because the colours of the two are superficially very alike. Looking at them as they are seen fairly close together in the sky, it requires a conscious mental effort to realize that Mars is one of the smaller planets within our Solar System, while Antares is a red supergiant star, far superior to our Sun in terms of both size and luminosity. This will be even harder to credit towards the end of 2005, when Mars will be so spectacular – though by then it will, of course, have moved well away from Antares. The two objects are well worth looking at this month, though it means getting up early; Antares and Mars can be seen low in the south-east an hour or so before dawn.

*The Colours of the Planets.* All the bright planets are visible this month at various times during the night, so their colours may be compared. Indeed, it is these colours that led, in part at least, to the names of the planets. There is certainly no doubt about Mars. Its redness is striking even when, as at present, it is relatively distant from the Earth, and is not as bright as the average first-magnitude star. It is easy

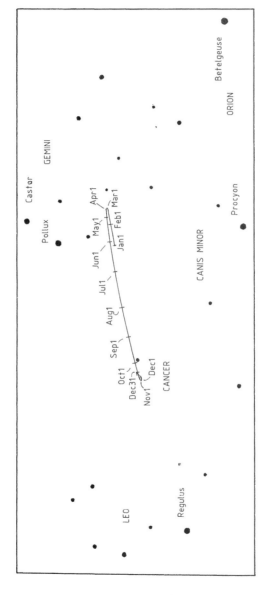

**Figure 1.** The path of Saturn against the stars of Gemini and Cancer during 2005.

to understand why the ancients named it after the God of War; red suggests blood. The red hue of Mars is due to the surface material; the Martian dust is reddish-brown, because it contains weathered limonite, a brown iron oxide, and smaller amounts of red haematite, one of the main ingredients of rust on Earth.

Mercury, elusive and quick-moving, was appropriately named in honour of the Messenger of the Gods. When visible with the naked eye it is always low over the horizon, and so it seems to twinkle obviously (the popular notion that planets can never twinkle is wrong, though it is true that a planet, which shows up as a small disk, scintillates less than a star, which is to all intents and purposes a point source). It is said that Mercury is pinkish, but the colour is not pronounced even when the planet is seen through a telescope.

Venus is creamy-white, and is the most beautiful of the planets when seen with the naked eye. Certainly it is well named after the Goddess of Beauty. Telescopically, Venus is a disappointment, but the slightly creamy or off-white hue can often be noticeable, even though when well above the horizon the disk may appear colourless.

The giant outer planets, Jupiter and Saturn, are both strongly yellowish. In the case of Jupiter, the hue is not particularly noticeable with the naked eye, but Saturn looks decidedly 'leaden', and was regarded as baleful; the ancients could have had no idea that Saturn is in fact the least dense and the most beautiful of all the members of the Sun's family. The rings are more reflective than the disk, and so at opposition this month, with the rings splendidly displayed, the planet will be bright, about magnitude −0.3. Saturn is still very well placed in the northern hemisphere of the sky, and so will be a fine sight for observers in Europe and the United States.

Of the three telescopic planets, Uranus (dimly visible with the naked eye when at its best) is greenish and Neptune has a bluish cast. Pluto is said to be rather yellow, but it is so faint that ordinary-sized telescopes used by amateurs will show it only as a colourless, star-like point.

Incidentally, the names we use for the planets are Roman, though the gods and goddesses themselves are Greek (when Rome became dominant, the Greek deities were simply taken over, though sometimes with minor alterations). The Greek names still survive in astronomy: for instance, 'Martian geography' is known as areography (Greek, Ares – the war god), while with Jupiter we still meet with the term 'zenocentric', from the Greek name Zeus – the ruler of Mount Olympus.

*Naked-eye Pleiades.* The Pleiades star cluster (Figure 2) is well placed this month. It is one of the most famous features of the sky and is associated with many myths and legends. The brightest star is Eta Tauri, Alcyone (magnitude 2.9); then come Atlas (3.6), Electra (3.7), Maia (3.9), Merope (4.2), Taygete (4.3), Pleione (5.1, variable), Celaeno (5.4) and Asterope (5.6).

How many separate stars in the Pleiades can you see with the naked eye when the sky is clear and dark? The popular nickname for the cluster is the 'Seven Sisters', and certainly people with average eyesight can see seven stars, but keen-sighted people can do much better than this. Asterope is on the fringe of naked eye visibility, and Atlas and Pleione are so close together that binoculars are needed to separate them well. If you can see a dozen stars you are doing very well. The record, held by the nineteenth-century German astronomer Edward Heis, is said to be nineteen, without optical aid. There are about 500 stars in the Pleiades cluster altogether.

**Figure 2.** This composite image of the Pleiades star cluster was taken by the Palomar 48-inch Schmidt telescope. The image is from the second Palomar Observatory Sky Survey, and is part of the Digitized Sky Survey. The separate images were taken between 5 November 1986 and 11 September 1996. (Image courtesy of NASA, ESA and AURA/Caltech.)

# February

MERCURY passes through superior conjunction on 14 February, and remains unsuitably placed for observation throughout the month.

VENUS, magnitude −3.9, continues to move closer to the Sun, though observers in Mediterranean latitudes and further south will still be able to see it for a short while low in the eastern sky before dawn.

MARS continues to be visible as a morning object in the south-eastern sky, magnitude +1.3. Unfortunately for observers in the latitudes of the British Isles it reaches its maximum southerly declination of almost 24° this month so that for them it will be very poorly placed for observation. Only under unusually clear conditions are they likely to glimpse the planet, and then only for a short while before the morning twilight inhibits observation. Mars continues to move eastwards in the constellation of Sagittarius.

JUPITER continues to be visible as a brilliant object in the morning skies, rising above the east-south-eastern horizon well before midnight. Its magnitude is −2.3. The planet reaches its first stationary point on 2 February, before turning retrograde. Jupiter remains in Virgo, west of Spica, during February. Figure 3 shows the path of Jupiter against the constellations of Virgo and Libra during 2005.

SATURN, magnitude −0.2, continues to be visible as a bright object in the evenings in the south-western quadrant of the sky. It is still retrograding slowly in the constellation of Gemini, south of Castor and Pollux. The rings of Saturn present a beautiful spectacle to the observer with a small telescope. The diameter of the minor axis of the rings is now 18 arcseconds, almost exactly the same as the polar diameter of the planet itself. The rings were last at their maximum opening in 2002. They will next appear edge-on in 2009.

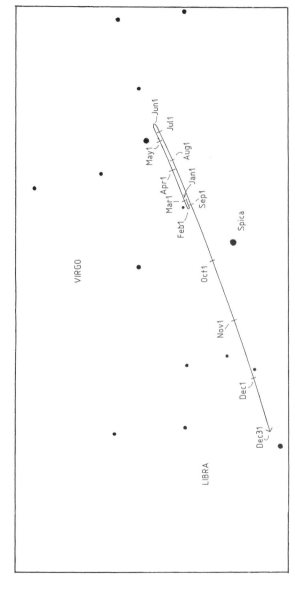

**Figure 3.** The path of Jupiter against the stars of Virgo and Libra during 2005.

*The Twinkling of Sirius.* On February evenings the brilliant star Sirius is well placed for observation. Observers in northern temperate latitudes should look for it in the south, rather low down. Though it is actually a white star of spectral class A, it seems to twinkle violently, flashing all the colours of the rainbow. Twinkling, or scintillation, has nothing to do with the stars themselves. It is due entirely to the unsteady atmosphere of the Earth. When a star is low, its light comes to the observer after having passed through a deep layer of atmosphere, so that the twinkling is increased. The effect is more noticeable with Sirius than with any other star, simply because Sirius is so brilliant.

It is sometimes said that stars twinkle, while planets do not. This is not entirely true; when a bright planet is seen low in the sky, it will twinkle quite noticeably except when the air is very calm and steady. However, a planet twinkles less than a star because it appears as a small disk, whereas a star is effectively a point source of light.

*Saturn Seen With Binoculars.* Saturn continues to be a conspicuous object this month. The ring system is a beautiful sight, and when at opposition in January the planet is brighter then any star apart from Sirius and Canopus. Unfortunately, the rings cannot be properly seen without the aid of a telescope. Low-power binoculars will do no more than show that Saturn is a disk rather than a point of light. With ×12 or higher magnification, it is just possible to discern that there is something unusual about the shape, particularly when the rings are almost fully open as they are throughout 2005, but there is no chance of seeing that the rings are responsible.

Titan, Saturn's largest satellite, is comparable in size with the planet Mercury, and is just below magnitude 8. It should, in theory, be visible with binoculars, although its angular distance from Saturn is usually little more than 3 arcminutes; it is going to be difficult, but it is always worth a try when Titan is near greatest elongation.

*Obscure Constellations.* The evening skies this month are still graced by the presence of Orion and his retinue, but there are also some very obscure groups on view: Lynx, the Lynx, and Camelopardalis, the Giraffe, are high in the sky, and Hydra, the Watersnake, with its companions in the south. Oddly enough, not all of these obscure constellations are modern inventions. Hydra was one of the original forty-eight constellations listed by Ptolemy of Alexandria, the last great astronomer of Classical times, who drew up his star catalogue (based on the earlier work of Hipparchus) in the second century AD. Hydra has the distinction of being the largest constellation in the sky, but it is also one of the dullest, since the reddish, second-magnitude Alphard, the 'solitary one', is its only bright star.

Cancer, the Crab, lying in the triangle defined by Regulus, Procyon and the Twins of Gemini, and also well placed this month, is another obscure constellation included in Ptolemy's list; it is notable only because it contains the naked-eye open cluster Praesepe, and because it is a member of the Zodiac. In shape, it resembles a very dim upside-down letter 'Y'. Its neighbour to the north, Lynx, and the adjoining pattern of Camelopardalis, were added to the sky map by Hevelius of Danzig in 1690, but with little justification, since they are barren of both bright stars and interesting objects. There is even less to be said for two more of Hevelius's additions, Leo Minor, the Little Lion, and Sextans, the Sextant, which border Leo. On the other hand, Monoceros, the Unicorn, also formed by Hevelius, is of some note. It lies in the area enclosed by imaginary lines joining Sirius, Betelgeux (or Betelgeuse) and Procyon, and is rich in star fields and telescopic objects, since the Milky Way runs right through it.

Of other obscure constellations, the Zodiacal patterns of Capricornus, the Sea Goat, Aquarius, the Water-bearer, and Pisces, the Fishes, were listed by Ptolemy. So, rather surprisingly, were several small groups in the region of Cygnus, namely Sagitta, the Arrow, Delphinus, the Dolphin, and Equuleus, the Little Horse; Equuleus consists only of a small, dim triangle of stars. Also ancient are Corvus, the Crow, and Crater, the Cup, which adjoin Hydra, and are well seen in late winter and early spring. Corvus is made up of a quadrilateral of moderately bright stars, and can be quite prominent, but Crater is very faint, and lacks any definite shape.

For large numbers of obscure constellations, one must, however, turn to the far southern sky. Near the southern celestial pole, and so invisible from European latitudes, are groups such as Octans, the Octant, Mensa, the Table – originally Mons Mensae, the Table Mountain – and Horologium, the Clock, which, with others of their kind, seem not to merit separate existence. However, it is most unlikely that any further revisions to the list of eighty-eight officially recognized constellations will be made now.

# March

*New Moon:* 10 March                    *Full Moon:* 25 March

*Equinox:* 20 March

*Summer Time* in the United Kingdom commences on 27 March.

MERCURY attains greatest eastern elongation (18°) on 12 March and is visible as an evening object until the middle of the month for observers in equatorial and northern latitudes. For observers in northern temperate latitudes, this will be the most favourable evening apparition of the year. Figure 4 shows, for observers in latitude 52°N, the changes in azimuth (true bearing from the north through east, south and west) and altitude of Mercury on successive evenings when the Sun is 6° below the horizon. This condition is known as the end of evening civil twilight and in this latitude and at this time of year occurs

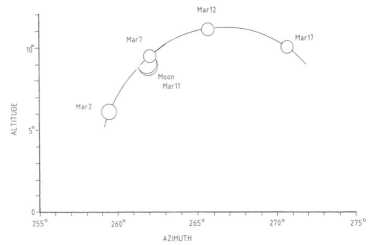

**Figure 4.** Evening apparition of Mercury, from latitude 52°N. (Angular diameters of Mercury and the crescent Moon not to scale.)

about thirty-five minutes after sunset. The changes in the brightness of the planet are indicated by the relative sizes of the circles marking Mercury's position at five-day intervals. At the beginning of this period of visibility its magnitude is −1.2, but this has faded to +1.0 by mid March.

**VENUS**, for observers in equatorial and southern latitudes, continues to be visible for a short while, low in the eastern morning sky before dawn, but only for the first fortnight of March. Its magnitude is −3.9. Venus passes slowly through superior conjunction on the last day of the month. It is not suitably placed for observation by those in the latitudes of the British Isles.

**MARS**, magnitude +1.1, continues to be visible in the south-eastern quadrant of the sky in the early mornings for some time before twilight renders observation impossible. It remains a very difficult object to detect for those observers in the latitudes of the British Isles. During the month, Mars moves from Sagittarius into Capricornus.

**JUPITER** is still visible as a brilliant object in the night sky, magnitude −2.4. By the end of the month it may be seen rising above the east-south-eastern horizon soon after sunset.

**SATURN**, magnitude 0.0, continues to be visible as an evening object in the south-western sky. On 22 March, it reaches its second stationary point and resumes its direct motion, in the constellation of Gemini.

*The Discovery of Pluto.* Seventy-five years ago this month, on 13 March 1930, the discovery of the ninth planet, Pluto, was announced. The planet had been tracked down by a 24-year-old farmer's son named Clyde Tombaugh, following a painstaking photographic search. Clyde Tombaugh was born on 4 February 1906 at Streator, Illinois, where his family ran a farm. He was the eldest of six children; his family could not afford to send him to college, and he cycled to school, eleven kilometres (seven miles) and back each day. By the age of eleven he had already become a competent farm worker, but his interests were wide-ranging. History and geography fascinated him, and in 1918 a first view through his uncle's telescope turned his attention to astronomy.

In 1922 the Tombaugh family moved to a farm in Kansas, and it was here that Clyde made his first telescope. It had an 8-inch mirror, a wooden tube and a rudimentary mounting, but it worked well. Later telescopes led him to make systematic observations, particularly of the planet Mars. Somewhat hesitantly, he sent some of his Mars drawings to Vesto M. Slipher, Director of the Lowell Observatory at Flagstaff in Arizona, and Slipher was impressed by the young man's talent. The drawings had arrived at just the right time because Slipher was looking for an assistant to carry out a search for a new planet, and it seemed that Tombaugh might be a suitable candidate.

The eighth planet, Neptune, had been tracked down in 1846 because of its gravitational perturbations on the motion of Uranus. Yet there was still something 'not quite right' about the movements of the outer planets, and Percival Lowell, founder of the observatory at Flagstaff, had concluded that there was another planet awaiting discovery. He worked out a position for his 'Planet X' and searched, but to no avail, and after his death in 1916 nothing more was done for some years – at least at Flagstaff. However, abortive searches were carried out elsewhere.

In 1928, Slipher decided to make a fresh attempt, and acquired a fine 13-inch refracting telescope especially for the purpose. The search was entrusted to Clyde Tombaugh, and in 1929 he arrived at the Observatory to begin work. The method was photographic. As we know, stars are so remote that their individual or 'proper' motions are very small, but a nearby object such as a planet will shift noticeably against the starry background after even a night or two. Tombaugh began a systematic hunt. If two photographs of the same region of the sky were taken over an interval of several nights, and then compared using a device known as a Blink-Comparator, a planet would betray itself by its motion.

It was laborious work, but Tombaugh was both patient and skilful, and success was not long in coming. Images taken on 23 and 29 January 1930 (Figure 5) showed a dot of light that moved in just the expected manner, and the hunt was over. Tombaugh examined the plates on 18 February, and saw that the object had to be moving well beyond the orbit of Neptune. In his own words:

> I walked down the hall to V.M. Slipher's office. Trying to control myself, I stepped into his office as nonchalantly as possible. He looked up from his desk work. 'Dr. Slipher,' I said, 'I have found your Planet X.' I had never come to report a mistaken planet suspect.

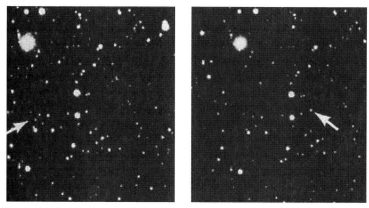

**Figure 5.** Discovery images of Pluto acquired by Clyde Tombaugh using the 13-inch refractor at the Lowell Observatory on 23 January (left) and 29 January 1930. The position of Pluto is shown by the white arrows, and its motion during the six days between the two photographs is obvious. (Images courtesy of Lowell Observatory, Flagstaff, Arizona.)

He rose right up from his chair with an expression in his face of both elation and reservation. I said, 'I'll show you the evidence.'

I explained that I had measured the shift to be consistent on the three plates, and that all the images were in the correct positions. Slipher kept flicking the shutter back and forth, studying the images. Then I said, 'The shift in my opinion indicates that the object is well beyond the orbit of Neptune.'

Then Slipher said, 'Don't tell any one until we follow it for a few weeks. This could be very hot news.' The excitement was intense. Another era for the Lowell Observatory was suddenly ushered in. The announcement three weeks later would cause excitement all over the world.

It did! The announcement was made on 13 March, Percival Lowell's birthday, and exactly 149 years after Herschel had identified the seventh planet, Uranus. Suddenly Clyde Tombaugh became world-famous. Examination of earlier plates showed that Pluto had been recorded twice in 1915, but had been missed because it was fainter than expected. However, failure to detect the planet in 1919 had been down to sheer bad luck. When the plates were re-examined, it was found that Pluto had been recorded twice – but once the image fell upon a flaw in the plate and on a second occasion Pluto was masked by an adjacent bright star.

Clearly the planet had to be named, and many suggestions were made, including Odin, Persephone, Chaos, Atlas, Tempus, Lowell, Minerva, Hercules, Daisy, Pax, Newton, Freya, Constance and Tantalus. Eventually, the name 'Pluto', suggested by an 11-year-old Oxford girl, Venetia Burney, was the favoured choice, and it was appropriate enough; Pluto was the God of the Underworld and the planet named in his honour must be a decidedly gloomy place.

In fact, Pluto proved to be an enigma. Its highly eccentric orbit, which swings it inside that of Neptune, was unlike that of any other planet. Moreover, Pluto turned out to be very small and is now known to be smaller even than the Moon. This means that it is of very low mass, and could not possibly exert any measurable perturbations on giant planets such as Uranus and Neptune. Either Lowell's reasonably accurate prediction was sheer luck, or else the real Planet X remains to be discovered. But this in no way detracts from Clyde Tombaugh's achievement. Pluto was at that time below fourteenth magnitude, and could easily have been overlooked.

*Jules Verne.* The famous French writer Jules Gabriel Verne, the so-called 'Father of Science Fiction', died on 24 March 1905, one hundred years ago this month. Born in Nantes on 8 February 1828, the eldest son of a lawyer, Jules Verne initially studied law, and gained a law degree, but while still a student he began writing, publishing a number of plays and some short stories. He married a wealthy young widow in 1857, and after several trips abroad he indulged his passion for writing. In 1863, a small book entitled *Five Weeks in a Balloon* appeared; it read like an authentic travel diary, but the adventures seemed fantastic. The book launched a new type of writing – science fiction. Its author was Jules Verne.

By the time Verne died, he had written well over seventy novels. They were well researched and were as scientifically accurate as was possible for the time. Verne's most famous works include *Journey to the Centre of the Earth, From the Earth to the Moon, Around the Moon, Twenty Thousand Leagues under the Sea* and *Around the World in Eighty Days.* Verne was an author well ahead of his time. He predicted submarines, flying machines, skyscrapers and even the Moon landings. His works of fiction inspired some of the world's most influential scientists, including three pioneers of modern rocketry and spaceflight – Konstantin Tsiolkovsky, Robert Goddard and Hermann Oberth.

The resemblance of Verne's novel *From the Earth to the Moon* to the actual Apollo Moon flights is uncanny. Three space pioneers (the same number of astronauts as the crew of an Apollo spacecraft) are shot into space from an enormous cannon located in Florida, USA (not far from where the Kennedy Space Center would later be built). Verne realized that his readers would understand a giant cannon better than a rocket. Of course, the sudden jolt as the cannon fired would have killed the would-be astronauts, but Verne was correct in that he knew the space capsule would have to reach a certain critical speed (escape velocity) to escape from the pull of Earth's gravity and reach the Moon. *From the Earth to the Moon* ended with the successful launch of the craft, but readers would have to wait four years for the sequel, *Around the Moon*, to find out what happened. At the end of the journey, when the three astronauts return to Earth, they splash down in the Pacific Ocean – another similarity to the real Apollo flights.

Verne continued to work and produce novels right up to his death at the age of 77. His books continue to entertain and inspire people today, one hundred years after his death. They will surely do so for many years to come.

# April

*New Moon:* 8 April                    *Full Moon:* 24 April

MERCURY is visible as a morning object after the middle of the month, though not to observers further north than Mediterranean latitudes. It passes through aphelion on 21 April and reaches greatest western elongation (27°) only five days later. For observers in southern latitudes this will be the most favourable morning apparition of the year. Figure 6 shows, for observers in latitude 35°S, the changes in azimuth (true bearing from the north through east, south and west) and altitude of Mercury on successive evenings when the Sun is 6° below the horizon. This condition is known as the beginning of morning civil twilight and in this latitude and at this time of year occurs about thirty minutes before sunrise. The changes in the brightness of the planet are indicated by the relative sizes of the circles marking Mercury's position at five-day intervals. It will be noticed that Mercury is at its brightest after it reaches greatest western elongation on 26 April. During the month its magnitude brightens from +1.2 to +0.3.

VENUS, having passed through superior conjunction at the end of last month, remains unsuitably placed for observation until almost the end of the month. However, observers in the northern temperate latitudes may be able to glimpse it during the last week of April, low above the western horizon, just after sunset. Its magnitude is −3.9.

MARS continues to be visible as a morning object in the south-eastern sky for a while before sunrise. Its magnitude brightens slowly during the month from +0.9 to +0.6 as it passes from Capricornus into Aquarius. Around the middle of the month Mars can be used as a guide to locating Neptune since Mars passes 1.2° south of Neptune on 13–14 April. Observers should refer to the chart for Neptune (Figure 16) given with the notes for August. The rectangular box in Figure 9, given with the notes for May, indicates the area of the Neptune diagram.

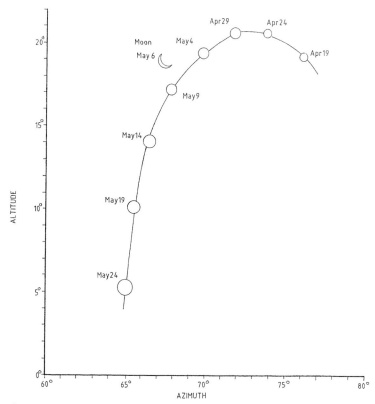

**Figure 6.** Morning apparition of Mercury, from latitude 35°S. (Angular diameters of Mercury and the crescent Moon not to scale.)

**JUPITER,** magnitude −2.5, is visible throughout the hours of darkness since it comes to opposition on 3 April. It is moving slowly retrograde in the constellation of Virgo. Jupiter's least distance from the Earth is 667 million kilometres (414 million miles).

**SATURN** continues to be visible as an evening object in the constellation of Gemini. Its magnitude is +0.2.

*Ursa Major at the Zenith.* There can be few people who cannot recognize Ursa Major, the Great Bear. With its seven leading stars making up the familiar outline usually nicknamed the 'Plough' in

Britain and the 'Big Dipper' in North America, it cannot be mistaken. During April evenings, it is worth noting that, as seen from northern Europe, Canada and the northern United States, the pattern is practically overhead, making it even easier to identify. It lies not far from the north celestial pole, so that from cities such as London or New York it is circumpolar; it can always be seen somewhere whenever the sky is sufficiently clear and dark.

To be accurate, Ursa Major contains many stars as well as those of the Plough. It is a large constellation, and there are some interesting objects in it, including M97, the Owl planetary nebula (unfortunately much too faint to be seen with small telescopes), and some interesting galaxies, including the spiral M81 and the irregular M82, which is a well-known radio source.

The seven Plough stars (Figure 7) are Alpha (Dubhe), Beta (Merak), Gamma (Phad or Phekda), Delta (Megrez), Epsilon (Alioth), Zeta (Mizar) and Eta (Alkaid or Benetnasch). Of these, five make up a moving cluster whose members share a common motion through space. The two unconnected stars are Alkaid and Dubhe, which are moving across the sky in a direction opposite to that of the remaining five. Over a sufficiently long period of time, the Plough will lose its familiar shape. All the Plough stars are white (spectral type A or B) with

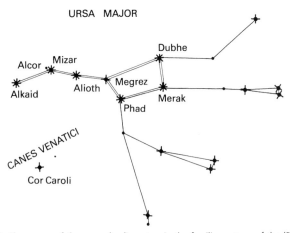

**Figure 7.** The names of the seven leading stars in the familiar pattern of the 'Plough' or 'Big Dipper', the most readily identifiable stars in the large constellation of Ursa Major, the Great Bear.

the exception of Dubhe, which is of spectral type K, and is distinctly orange when viewed through binoculars or a telescope. Merak and Dubhe are known as the 'Pointers', because they show the way to the Pole Star, Polaris.

Ancient astronomers apparently rated Megrez as equal to the other stars in the pattern, but it is now almost a magnitude fainter than Phad. Either there has been a real fading or, more likely, there has been an error in recording or interpretation. However, Megrez has been suspected of slight variability in modern times.

Mizar is probably the most famous double star in the sky. It forms a naked-eye pair with Alcor, and with a small telescope Mizar itself is seen to be made up of two rather unequal components. It is a binary, but the period of revolution is extremely long.

Ursa Major contains no star of the first magnitude: the brightest are Alioth and Dubhe (magnitude 1.8) and Alkaid (magnitude 1.9), but it is as easily recognized as any constellation in the sky.

*An Unusual Solar Eclipse.* The eclipse of the Sun on 8 April will be seen from the Pacific Ocean, parts of Central America and the northern part of South America. Nothing of it will be visible from Europe. However, this eclipse is an example of one of the rarest types of solar eclipse – an annular-total eclipse, also called a 'hybrid' eclipse. In the hundred years from 2001 to 2100, there are only seven such hybrid eclipses. On these rare occasions, the Sun and Moon appear almost *exactly* the same size in the sky. Consequently, an annular eclipse will be visible from both ends of the track, while a total eclipse will be seen around the mid-point of the eclipse path, where the cone of the Moon's umbral shadow just reaches the Earth's surface (Figure 8). The width of the eclipse path in an annular-total eclipse changes markedly along the track: first it narrows down, then the width increases again, and then it grows narrow once more. These variations are due to the eclipse changing from an annular eclipse to a total, then back to an annular again.

Totality on 8 April will be brief, lasting just 42 seconds, and will be visible only far out at sea in the South Pacific. The Sun should be approaching the minimum of its roughly eleven-year solar cycle and so the corona during totality should be of the 'minimum' variety – though in recent years the Sun has been behaving in a rather unexpected manner, and we can never be sure just what the corona will be like. Having changed back into an annular eclipse, the eclipse track first

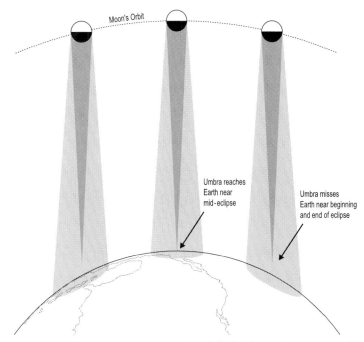

**Figure 8.** An annular-total eclipse is the rarest kind of solar eclipse. The narrow cone of the Moon's umbral shadow just reaches the Earth's surface near the middle of the eclipse track (producing a short total eclipse), but because of the curvature of the Earth it does not reach the surface towards both ends of the track (producing only an annular eclipse).

strikes land in Central America, then crosses northern Colombia and ends in northern Venezuela at sunset. The North Island of New Zealand will see a partial eclipse at sunrise, and much of western South America, Central America and southern parts of the USA will see a partial eclipse in the afternoon.

# May

**MERCURY** is not visible to observers in the latitudes of the British Isles because of the long duration of morning twilight. However, it may be seen by those observers further south where it remains visible as a morning object, except for the last week of the month. Those observers should refer to the diagram given with the notes for April.

**VENUS**, magnitude −3.9, is slowly beginning to move out of the long evening twilight, becoming a brilliant object, though only low in the western sky for a short while after sunset.

**MARS**, its magnitude brightening from +0.6 to +0.3 during the month, is still visible as a morning object in the south-eastern quadrant of the sky. The planet is in the constellation of Aquarius. Figure 9 shows the path of Mars against the constellations of Sagittarius, Capricornus and Aquarius during the early months of the year. Mars can be a useful guide to locating Uranus around the middle of the month as Mars passes 1.1° south of Uranus on 15–16 May. Observers should refer to the chart for Uranus (Figure 18) given with the notes for September. The rectangular box in Figure 14, given with the notes for July, indicates the area of the Uranus diagram.

**JUPITER** continues to be visible as a brilliant object in the evening sky, in the constellation of Virgo. Its magnitude is −2.3. The four Galilean satellites of the planet are readily observable with a small telescope or even a good pair of binoculars provided that they are held rigidly.

**SATURN**, magnitude +0.3, is still visible in the western sky in the evenings, setting around four hours after sunset. Saturn passes 7° south of Pollux, in the constellation of Gemini, at the end of the month.

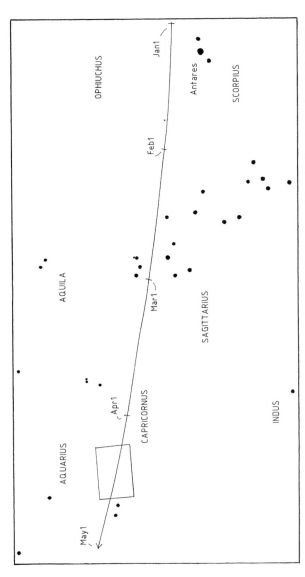

**Figure 9.** The path of Mars against the stars of Scorpius, Sagittarius, Capricornus and Aquarius from January to early May 2005. The rectangular box indicates the area of the Neptune diagram (Figure 16) included with the notes for August.

*Arcturus.* Alpha Boötis, or Arcturus, the leader of Boötes, the Herds-man, is very prominent this month. Its declination is +19°, so that it can be seen from every inhabited country, though from the southernmost part of New Zealand it is always low down. It is the fourth brightest star in the sky (third, if you do not combine the two components of Alpha Centauri), and the brightest in the Northern hemisphere of the sky. There are four stars that are almost equal: Arcturus (magnitude −0.04), Vega (+0.03), Capella (+0.08) and Rigel (+0.12). Vega is bluish, Rigel is pure white, Capella yellowish and Arcturus orange. The latest measure-ments indicate that Arcturus is thirty-seven light years away, and is 115 times as luminous as the Sun, though admittedly this does not seem much in comparison with a celestial 'searchlight' such as Rigel – in fact, Rigel is around 500 times as luminous as Arcturus.

There are few definite mythological legends associated with Boötes. It has been said that the herdsman was honoured with a place in the heavens because he invented the plough drawn by two oxen. Arcturus is most easily found by extending the curve of the three stars in the tail of the Great Bear (Figure 10). Indeed, the star is said to take its name from its proximity to the stars of the Great Bear and Little Bear; its name in

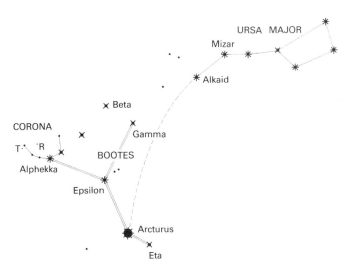

**Figure 10.** The brilliant orange star Arcturus in Boötes, the Herdsman, is most easily found by following round the curve of the three stars in the tail of Ursa Major (also the handle of the Plough), as shown by the dotted line here.

Greek means 'guardian of the bears' or 'bear watcher'. For some reason or other seamen of ancient times regarded Arcturus as unlucky, and the Roman writer Pliny even refers to it as 'horridum sidus' – which is indeed strange, because Arcturus is such a beautiful star.

The stars are so far away that their individual or proper motions are very slight, and the constellations look virtually the same now as they must have done to King Canute, Julius Caesar or the builders of the Pyramids. But they do creep along compared with each other, and Arcturus, which is one of our nearer neighbours, has a proper motion of 2.3 arcseconds per year. Stellar proper motion was discovered in 1718 by Edmond Halley. He realized that the stars Arcturus, Sirius and Aldebaran had shifted appreciably from the positions recorded by the famous Greek astronomer Hipparchus around 1,800 years earlier. A large proper motion is usually an indication that a star lies relatively close to the Sun, and for this reason Arcturus was once thought to be the closest of the bright stars. This is quite wrong: of the first-magnitude stars, Alpha Centauri, Sirius, Vega, Procyon, Altair, Pollux and Fomalhaut are all closer. However, of the very brightest stars only Alpha Centauri has a greater proper motion than Arcturus: 3.7 arcseconds per year.

Arcturus appears to be fast moving because it probably belongs to an older population of stars – known as Population II stars – located within the thick disk of our Galaxy, the Milky Way. The thick disk is generally formed of old stars that lie well above or below the galactic plane, unlike thin-disk stars, such as our Sun, which lie close to the plane. However, thick-disk stars tend to move in highly inclined and elliptical orbits around the galactic core, so Arcturus is travelling along a highly inclined orbit cutting through the galactic plane, which gives it a velocity relative to the Sun that is higher than the other bright stars. Compared with the other surrounding stars, which orbit the Galaxy in roughly circular orbits, Arcturus falls behind by around a hundred kilometres per second (about sixty miles per second).

Arcturus must have reached naked-eye visibility half a million years ago, and has since brightened steadily, and is now just about at its nearest to the Sun. In the future it will draw away, until it drops below naked-eye visibility in about half a million years from now.

*Coma and Canes Venatici.* South of the Plough, and well placed on May evenings, lies the constellation of Coma Berenices, or Berenice's

Hair. An intriguing old legend is attached to it. It is said that when Ptolemy Euergetes, the warrior King of Egypt, set out on a dangerous campaign against the Assyrians, his wife Queen Berenice vowed to dedicate her beautiful hair to the gods if her husband came back safely. On his return, she kept her promise, and the gleaming tresses were placed in the sky! Yet Coma is not an ancient constellation in its own right: it was added to the sky by Tycho Brahe during the latter part of the sixteenth century.

At first glance Coma gives the superficial impression of being a large, dim open star cluster. Its principal stars (Beta, Alpha and Gamma) are only of the fourth magnitude, but although the pattern has no bright stars, there are many faint ones, and there are also numerous telescopic galaxies. Alpha Comae makes a triangle with Eta Boötis and Epsilon Virginis, and is not hard to locate as there are no other naked-eye stars close to it. The globular cluster M53 is in the same low power field as Alpha. It is of magnitude 7.7, and is discernible with binoculars as a dim blur. Coma is also in the region of the northern galactic pole, where we look perpendicular to and above the galactic plane, and many faint external galaxies are visible. The Black-Eye galaxy, M64, magnitude 6.6, is within a degree of the star 35 Comae, not far from Alpha.

Between Coma and the Great Bear lies another small constellation, Canes Venatici (the Hunting Dogs), formed by Hevelius in 1690. The only bright star is Cor Caroli (Charles' Heart), of the third magnitude. The globular cluster M3, magnitude 6.4, lies at the extreme edge of the constellation, close to the border with Coma. M3 is almost midway between Cor Caroli and the brilliant Arcturus in Boötes. Canes Venatici is also crowded with galaxies. Probably the most celebrated of these is the Whirlpool Galaxy, M51 (Figure 11), magnitude 8.4, which lies quite close to Alkaid, the star at the end of the Great Bear's tail. M51 was the first galaxy to have its spiral structure recognized; it was in 1845 when the Third Earl of Rosse used his remarkable home-made 72-inch reflecting telescope at Birr Castle in Ireland to observe it. At the time, only the Birr telescope was capable of showing the forms of galaxies. It was a major breakthrough because they are now recognized as star systems in their own right, lying well beyond our own Milky Way galaxy.

**Figure 11.** The beautiful Whirlpool Galaxy, M51, in the constellation of Canes Venatici, comprising the large spiral galaxy NGC 5194 and its smaller companion NGC 5195, with which it is clearly interacting. This composite image was acquired with the 0.9-metre telescope at the Kitt Peak National Observatory. (Image courtesy of Todd Boroson/NOAO/AURA/NSF.)

# June

*Solstice:* 21 June

MERCURY is unsuitably placed for observers further north than Mediterranean latitudes. Further south it becomes visible after the middle of the month as an evening object low in the west-north-western sky at about the time of ending of evening civil twilight, its magnitude fading from −1.2 to +0.1 by the end of June.

VENUS, magnitude −3.9, is a brilliant object in the evening sky, though only visible above the western horizon for a while after sunset; in the latitudes of the British Isles, Venus sinks to only 5° above the horizon about thirty-five minutes after sunset.

MARS is still a bright object in the south-eastern quadrant of the sky in the mornings. During the month its magnitude brightens from +0.3 to 0.0. Mars commences the month in Aquarius, transits the extreme north-western portion of Cetus and ends the month in Pisces.

JUPITER is still visible in the evenings in the south-western sky in the constellation of Virgo. On 5 June it reaches its second stationary point, resuming its eastward motion towards Spica. By the end of June it is no longer visible after midnight. Its magnitude is −2.1.

SATURN is still visible low in the western sky in the early evenings, though observers in the latitudes of the British Isles will be unable to view it after the middle of the month because of the long duration of twilight. Its magnitude is +0.3.

PLUTO reaches opposition on 14 June, in the constellation of Serpens, at a distance of 4,482 million kilometres (2,785 million miles) from the Earth. A moderate-sized telescope is required to observe Pluto since its magnitude is +14.

*Comet 9P/Tempel 1 and Deep Impact.* Comet 9P/Tempel 1 is a typical short-period comet, and the intended target for NASA's Deep Impact spacecraft. Tempel 1 is on view this month in the region of Virgo, and should be easy to find with a small telescope or good binoculars; positions for it are given on page 142 of this *Yearbook*.

Comet Tempel 1 was discovered on 3 April 1867 by Ernst Tempel, an energetic comet hunter from Marseilles, France. The comet was well placed in 1867 thanks to a close approach to the Earth in May of that year. It was recognized as a short-period comet, with a period of 5.7 years, and was subsequently recovered in 1873 and 1879. Tempel 1 swung close to Jupiter in 1881, and its orbit was changed, making the comet an even fainter object. Subsequently, it was lost and was not definitely recovered until 1972, as the result of calculations by Brian Marsden. He found that further close approaches to Jupiter in 1941 and 1953 had decreased both the perihelion distance and the orbital period to values smaller than when the comet was first discovered. (One very diffuse single image was found on a photograph in 1967, but it did not provide definitive proof of the comet's recovery.) Since 1972 the comet has been seen at every return, and its orbit is well known.

Tempel 1 now has an orbital period of 5.5 years and a perihelion distance of roughly 1.5 AU; its orbit lies between the orbits of Mars and Jupiter. The comet's nucleus is probably elongated and about six kilometres (almost four miles) in diameter, but its size and shape are uncertain because we cannot see it. We have only imaged the nuclei of three comets – Halley, Borrelly and Wild 2. In July 2005, it is hoped to add a very detailed view of the nucleus of Tempel 1.

Due for launch in December 2004, the main Deep Impact spacecraft will carry a smaller 'impactor' spacecraft to Tempel 1 and release it into the comet's path for a planned collision. Images of the nucleus will be obtained as the craft approaches and tracks the comet. On 4 July 2005, the Deep Impact spacecraft will release the 370-kilogramme impactor on a course to hit the comet's sunlit side (Figure 12). A camera on the impactor will capture and relay images of the comet's nucleus seconds before the collision. The impact will blast a crater perhaps as large as a football stadium in the comet, ejecting ice and dust from the crater and revealing fresh material beneath. After release of the impactor,

**Figure 12.** An artist's impression of the Deep Impact spacecraft releasing the impactor, 24 hours before the impact event. Pictured from left to right are the target, comet Tempel 1, the impactor, and the fly-by spacecraft. (Image courtesy of NASA Jet Propulsion Laboratory, Univ. of Maryland and Ball Aerospace & Technologies Corp.)

the main fly-by spacecraft manoeuvres to a new path to observe the ejected material blasted outwards, and the structure and composition of the crater's interior. Meanwhile, on Earth, professional and amateur astronomers using large and small telescopes will observe the impact and its aftermath, including changes in the comet's activity. If everything goes according to plan, the Deep Impact mission should provide our very first glimpses inside a cometary nucleus.

*The Leaders of the Cross.* During June evenings, for southern observers, Crux Australis, the Southern Cross, is riding high in the south-west. Crux is the most famous pattern of the southern sky – and is, incidentally, the smallest constellation in the entire sky. Before Royer introduced it in 1679, the stars of the Cross had been included in neighbouring Centaurus (Figure 13).

We have to admit that the shape of Crux is rather more like a kite than an 'X', but the three leading stars are unmistakeable: Alpha, Beta and Gamma are of magnitude 0.8, 1.3 and 1.6 respectively, while the fourth, Delta Crucis, is of magnitude 2.8. A fifth and fainter star, Epsilon (magnitude 3.6), lies between Alpha and Delta.

Three of the leaders are hot and white or bluish, with B-type spectra; the fourth member of the main pattern, Gamma Crucis, is an orange-red giant of spectral type M. Alpha Crucis, or Acrux, is a splendid binary, magnitudes 1.4 and 1.9, with a separation of 4 arcseconds and visible with any small telescope.

As well as its brilliant stars, Crux contains the glorious open cluster NGC 4755, popularly known as the 'Jewel Box', and also the dark nebula known as the 'Coal Sack'. The Jewel Box cluster, around Kappa Crucis, lies not far from Beta; the brightest stars are blue or bluish-white, but there is one prominent red giant in their midst, providing a beautiful colour contrast. Close by is the Coal Sack, a dark nebula 60–70 light years across, which shows up clearly because the dust it contains blots out the light of stars in a rich part of the Milky Way beyond.

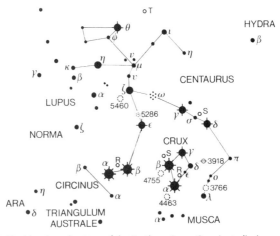

**Figure 13.** The kite-shaped pattern of the Southern Cross, Crux Australis, is surrounded on three sides by Centaurus, the constellation to which its stars once belonged. As well as its brilliant stars, Crux contains the glorious star cluster, the Jewel Box, NGC 4755, not far from Beta Crucis.

# July

EARTH is at aphelion (furthest from the Sun) on 5 July at a distance of 152 million kilometres (94 million miles).

MERCURY reaches greatest eastern elongation (26°) on 9 July. The long duration of twilight, together with its southerly declination, renders it unobservable to those in the latitudes of the British Isles. For observers further south it continues to be visible as an evening object in the south-western sky around the time of ending of evening civil twilight, until the middle of the month. During this period its magnitude fades from +0.2 to +1.0.

VENUS, magnitude −3.9, is a brilliant object in the evening sky, but for observers in the British Isles it is only visible low above the western horizon for about half an hour after sunset.

MARS continues to be visible as a morning object, though by the end of the month it should be possible to detect it low above the eastern horizon by midnight. Before dawn it will be seen fairly high on the meridian. Mars is still moving eastwards, but remains in the constellation of Pisces. During the month its magnitude brightens from −0.1 to −0.5. Figure 14 shows the path of Mars against the constellations during the rest of the year. The rectangular box indicates the area of the Uranus diagram (Figure 18), given with the notes for September.

JUPITER continues to be visible as a brilliant evening object in the south-western sky, magnitude −1.9.

SATURN is too close to the Sun for observation during this month as it passes through conjunction on 23 July.

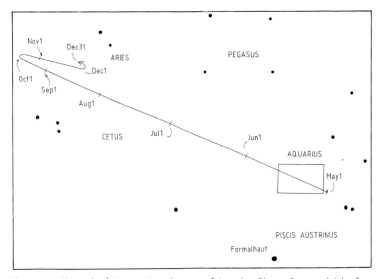

**Figure 14.** The path of Mars against the stars of Aquarius, Pisces, Cetus and Aries from May to December 2005. The rectangular box indicates the area of the Uranus diagram (Figure 18) included with the notes for September.

*Vega and Capella.* Throughout summer evenings, the southern aspect of the night sky is dominated by the so-called 'Summer Triangle', its corners marked by three first-magnitude stars, each the brightest star in its own constellation: Deneb in Cygnus, the Swan; Altair in Aquila, the Eagle; and Vega in Lyra, the Lyre or Harp. Of the three, Vega appears much the brightest and is decidedly bluish in colour. Next in brightness is Altair, which is off-white in colour, and finally there is Deneb, which is pure white. As these three stars appear to us, Vega is about twice as bright as Altair and more than three times brighter than Deneb. Vega is certainly more luminous than Altair, because Vega is further from us yet appears brighter: Altair is seventeen light years away with ten times the Sun's luminosity, while Vega is twenty-five light years distant and more than fifty times as luminous as the Sun. However, brilliant Vega is actually vastly inferior to Deneb, which is one of the most luminous supergiant stars known, but lies at a very great distance from us, and so appears fainter than the other two.

Vega is of magnitude +0.03 and is brighter than any other star visible from Britain apart from Sirius and Arcturus. During the late evenings in July, Vega is practically overhead. On the opposite side of the Pole lies Capella in Auriga, the Charioteer, whose magnitude of +0.08 means that it is only very slightly inferior to Vega, but it is of a completely different spectral type; its colour is yellow. Capella occupies the overhead position during winter evenings. Obviously, when Vega is high up, then Capella is low down and vice versa. Neither actually sets from the latitude of London but both become very low, and on July evenings Capella almost grazes the northern horizon.

Arcturus in Boötes, the Herdsman, is still on view, high in the south-western sky in the late evening. Here we have a star of yet another colour – a beautiful light-orange colour, contrasting sharply with the steely blue of Vega.

*Hercules and its Clusters.* This month the constellation of Hercules, commemorating the great hero of Greek mythology, is very high up as seen from northern latitudes, just to the west of the brilliant Vega; Hercules also attains a respectable altitude from countries well south of the equator. Although it is a sprawling pattern, Hercules is by no means a brilliant constellation, and there are only two stars above magnitude 3.0: Beta, or Kornephoros (magnitude 2.8), and Zeta, or Rutilicus (magnitude 2.8). The constellation's central region is known as the 'Keystone', because of its resemblance to the angular block found at the top of a masonry arch. The corners of the Keystone are marked by the four stars Eta (3.5), Zeta, Epsilon (3.9) and Pi Herculis (3.2), and make up the trapezium-shaped body of Hercules.

One of the most interesting stars in Hercules is Alpha, or Rasalgethi, whose name means literally 'the kneeler's head'. Alpha is a vast red supergiant, and lies well away from the main part of the constellation and near the brighter Rasalhague, or Alpha Ophiuchi (magnitude 2.1). Recent measures give the distance of Rasalgethi as 400 light years, with a luminosity about 500 times that of the Sun. It is cool, with a surface temperature of about 3,000 K, and its diameter is greater than that of the orbit of Mars. Rasalgethi is surrounded by a huge gaseous envelope, nearly 300 times the size of the star itself, consisting of material puffed off by the supergiant. It varies between magnitudes 3 and 4, with a very rough and inconstant period of a few months. Rasalgethi has a fifth-magnitude companion, 4.7 seconds of arc away, but the companion can be quite elusive, because it is overpowered by the glare of the primary. It is sometimes described as green in colour, but this is due mainly to contrast with the orange-red primary.

Another interesting object in Hercules is Zeta (Rutilicus), which is a fine binary; the components are of magnitude 2.9 and 5.5. The components can be seen separately with a moderate telescope, but the separation (currently 0.8 arcseconds) and position angle alter quickly, because the orbital period is only 34.5 years. At present, the pair is slowly widening.

Hercules contains two splendid globular clusters, M13 and M92 (Figure 15). M13 is the brightest of all the globulars, apart from the far southern Omega Centauri and 47 Tucanae, and the finest in northern skies. It is just visible with the naked eye under good conditions; binoculars show it well, and the outer parts at least can be resolved with a small telescope. It contains at least 100,000 stars; the diameter of the main cluster is over 150 light years, and the distance

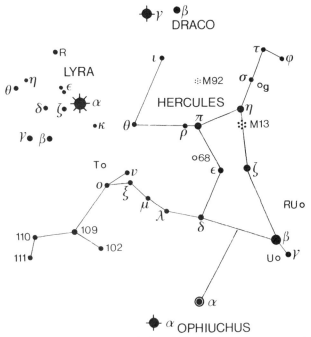

**Figure 15.** The stars of Hercules lie not far from the brilliant Vega. The positions of the two globular clusters in Hercules, M13, and the slightly fainter M92, are shown in the upper right quadrant of this map.

from us is around 25,000 light years. It is easy to locate, roughly midway between Zeta and Eta, on the western side of the Keystone.

In 1974, at the dedication of the Arecibo radio observatory in Puerto Rico, a radio message was beamed to M13 in the hope that some radio astronomer there will pick it up and reply. If this happens, readers of the *Yearbook* are urged to listen out for it around the year AD 52000!

M92 is slightly inferior to M13; it was discovered in 1888 by J.E. Bode. It can just be seen with the naked eye under the best possible conditions, and telescopically it looks much like M13. With low-power binoculars, M92 is in the same field of view as Iota Herculis and the Pi-Rho pair. M92 is about 27,000 light years distant, only slightly further away than M13; the diameter of the main cluster is over 100 light years. Precession means that around the year AD 14000, the cluster will be within one degree of the north celestial pole.

# August

*New Moon:* 5 August                    *Full Moon:* 19 August

MERCURY passes through inferior conjunction on 5 August and becomes visible as a morning object shortly after the middle of the month, visible in the eastern sky for a short while around the time of beginning of morning civil twilight. For Southern Hemisphere observers the visibility period is unlikely to be more than one week, while further north Mercury remains visible until beyond the end of the month. During the second part of August its magnitude brightens from +1.1 to −0.9. The planet attains greatest western elongation (18°) on 24 August.

VENUS, magnitude −4.0, continues to be visible as a brilliant object in the western sky after sunset. Disappointingly for observers in the latitudes of the British Isles, it is still only visible for about half an hour after sunset, low above the western horizon.

MARS is becoming quite a conspicuous object in the sky from the late evening onwards. During the month its magnitude brightens from −0.5 to −1.0. Mars is in the constellation of Aries.

JUPITER, magnitude −1.8, is still a bright evening object in the south-western skies. However, for observers in the latitudes of the British Isles it will only be visible for a short while before it sets, and by the end of the month it will be a difficult object to detect in the gathering twilight.

SATURN, magnitude +0.3, slowly becomes visible to observers in the latitudes of the British Isles during the second half of the month, low above the eastern horizon in the early mornings before it is lost in the morning twilight. Further south it is unlikely to be seen until the third week of August.

NEPTUNE is in the constellation of Capricornus and comes to opposition on 8 August. It is not visible to the naked eye since its magnitude is +7.8. The accompanying diagram (Figure 16) shows the path of Neptune against the stars during the year. The two brightest stars in the diagram are Theta Capricorni (magnitude +4.1, right ascension 21h 06.3m, declination −17.21°), and Iota Capricorni (magnitude +4.3, right ascension 21h 22.6m, declination −16.81°). To assist observers in locating this area it is shown as a rectangular box on the Mars diagram (Figure 9) given with the notes for May. At opposition Neptune is 4,345 million kilometres (2,700 million miles) from the Earth.

*Synodic Periods of the Planets.* The planet Neptune is at opposition this month; Uranus is at opposition early next month. During 2005 there are oppositions of Jupiter, Saturn, Uranus, Neptune and Pluto – as indeed there are in almost every year. With Mars, the situation is different. Mars is at opposition in November 2005, but there was no opposition in 2004 and there will not be one in 2006. This is because the synodic period of Mars is much longer than that of any other planet.

The synodic period is the mean interval between successive oppositions of a planet. To understand what is meant, let us first consider Pluto, which moves round the Sun at a mean distance of more than 5,900 million kilometres (over 3,600 million miles), though its orbit is unusually eccentric, and at its closest it can come within the orbit of Neptune. As well as having a large orbit, Pluto is slow-moving. In one year, the time taken for the Earth to go once round the Sun, Pluto covers only a tiny fraction of its orbit – so the Sun, Earth and Pluto are lined up every 366.73 days. Having been right round the Sun in 365.25 days, the Earth takes only an extra day and a half to catch up with Pluto.

With the closer planets, more time is required; the mean synodic periods for Neptune, Uranus, Saturn and Jupiter are 367.49, 369.66, 378.09 and 398.88 days respectively – so, for instance, oppositions of Jupiter occur approximately thirty-four days later in each year; the opposition of April 2005 follows that of March 2004, and there will also be oppositions in May 2006, June 2007, July 2008 and so on.

Mars is a special case. Its orbital velocity is comparable with that of the Earth, and the mean synodic period is therefore a great deal longer at 780 days. Consequently, oppositions of Mars do not occur every

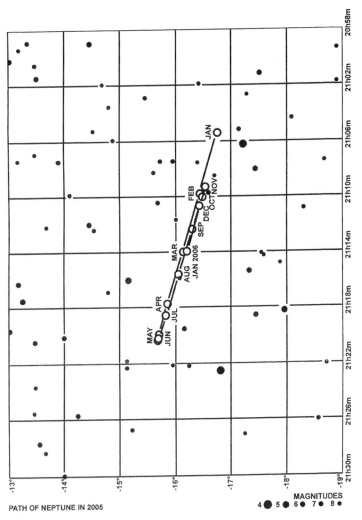

PATH OF NEPTUNE IN 2005

MAGNITUDES
4 ● 5 ● 6 ● 7 ● 8 ●

**Figure 16**. The path of Neptune against the stars of Capricornus during 2005.

year; there were oppositions in 2001 and 2003, and there is one in 2005, but there was none in 2004 – and following this year's opposition, 2006 will be another 'blank year' as far as Mars is concerned. Of course, Mars will not be invisible in 2006. At the start of the year it will not be long past opposition, and so will still be observable for much of the night. As the planet does not reach conjunction with the Sun until 23 October 2006, it will be visible for most of 2006, re-emerging in the morning sky just before dawn by the end of the year. However, it will be a long way away, and not even large telescopes will show much upon its tiny disk.

***Sagittarius and the Lettering of Stars.*** During August evenings, from northern latitudes, the constellation of Sagittarius, the Archer, may be seen low in the southern sky. From London or New York it is never visible to advantage, but from parts of the southern United States, such as California and Arizona, it is very prominent (Figure 17). It has no distinctive shape, but it contains various bright stars; eight of

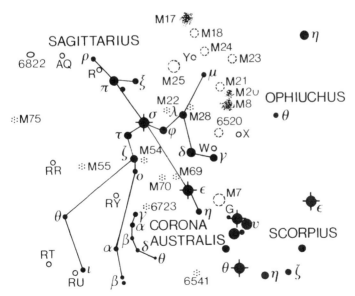

**Figure 17.** Sagittarius, the southernmost of the Zodiacal constellations, is not wholly visible from the United Kingdom. The centre of the Milky Way lies behind the star clouds here, the whole area is exceptionally rich, and it contains numerous Messier objects.

Sagittarius's brightest stars form a figure widely recognized as a 'teapot', with its handle to the east and its spout to the west. The Milky Way is very rich flowing through Sagittarius and, visually at least, the dense star clouds mask our view towards the centre of the Galaxy. The first-magnitude stars Deneb and Altair in the 'Summer Triangle' make almost a direct line to Sagittarius, with Altair in the middle position; this is probably the easiest way to find the constellation. Sagittarius is in the Zodiac, though it contains no planets at the present time; Mars passed through the pattern in February and March 2005.

When Bayer allotted Greek letters to the stars in each constellation, in his star maps of 1603, he usually followed the system of making the brightest star Alpha, the second brightest Beta, the third brightest Gamma, and so on. His system is convenient, and is still in use. The Greek alphabet is as follows:

| Alpha α | Epsilon ε | Iota ι | Nu ν | Rho ρ | Phi φ |
| Beta β | Zeta ζ | Kappa κ | Xi ξ | Sigma σ | Chi χ |
| Gamma γ | Eta η | Lambda λ | Omicron o | Tau τ | Psi φ |
| Delta δ | Theta θ | Mu μ | Pi π | Upsilon υ | Omega ω |

Perhaps, unfortunately, the strict brightness sequence is not always followed; thus in Orion, Beta (Rigel) is brighter than Alpha (Betelgeux or Betelgeuse). With Sagittarius, the alphabetical order can only be described as chaotic! The brightest stars are Epsilon (magnification 1.9), Sigma (2.0), Zeta (2.6), Delta (2.7), Lambda (2.8), Pi (2.9) and Gamma (3.0), with both Alpha and Beta considerably fainter. Incidentally, both Alpha and Beta (which is a wide, naked-eye double star) are too far south to rise in Britain or the northern United States.

# September

*New Moon:* 3 September          *Full Moon:* 18 September

*Equinox:* 22 September

MERCURY continues to be visible in the mornings for observers in tropical latitudes and in the Northern Hemisphere, but only for the first four or five days of the month, low above the east-north-eastern horizon, around the time of beginning of morning civil twilight. Its magnitude is −1.1. The planet passes through superior conjunction on 18 September.

VENUS continues to be visible as a magnificent evening object in the western sky after sunset, magnitude −4.1. Unfortunately, for observers in the latitudes of the British Isles, its increasing southerly declination means that it is visible for only about half an hour each evening after sunset, low above the south-western horizon.

MARS is now a very conspicuous object, as its magnitude brightens from −1.0 to −1.7 during September. It is becoming visible low in the eastern sky during the evening and remains visible for the rest of the night. Mars moves from Aries into Taurus during the month.

JUPITER, magnitude −1.7, is no longer visible to observers in the latitudes of the British Isles, but can still be seen by those in Mediterranean latitudes and further south. There it is visible low in the south-western sky for a short while after sunset.

SATURN continues to be visible as a morning object in the south-eastern quadrant of the sky. Its magnitude is +0.3. Saturn is in the constellation of Gemini.

URANUS, magnitude +5.7, is barely visible to the naked eye, though it is readily located with only small optical aid. Figure 18 shows the path

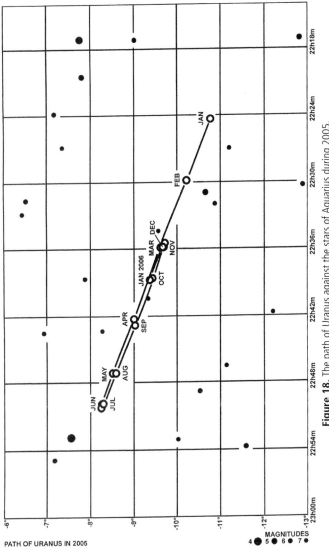

PATH OF URANUS IN 2005

MAGNITUDES
4● 5● 6● 7●

**Figure 18.** The path of Uranus against the stars of Aquarius during 2005.

of Uranus among the stars during the year. The two brightest stars in the diagram are Theta Aquarii (magnitude +4.2, right ascension 22h 17.1m, declination −7.76°) and Lambda Aquarii (magnitude +3.7, right ascension 22h 52.9, declination. −7.55°). To assist observers in locating this area it is shown as a rectangular box on the Mars diagram (Figure 14) given with the notes for July. At opposition on 1 September, Uranus is 2,851 million kilometres (1,772 million miles) from the Earth.

*The Names of Full Moons.* In the Northern Hemisphere, the full moon occurring closest to the autumnal equinox, which falls on 22 September this year, is known as the Harvest Moon. At this time of year, the retardation – that is to say, the time interval between moonrise on successive nights – is at its minimum; it may be no more than around fifteen minutes, compared with an hour or more at other times of the year. With the Moon rising only slightly later each evening, for the nights around Harvest Moon, this was said to be helpful to farmers gathering in their crops. Contrary to popular belief, the Harvest Moon looks the same as any other full moon.

Other full moons of the year have nicknames too, but only the Harvest Moon and the Hunter's Moon, which occurs a month later (usually in October), are well known. The North American periodical the *Farmers' Almanac* has listed the traditional names given to full moons. Apparently, many of these names date back to the Native Americans of what is now the northern and eastern USA. The various tribes kept track of the seasons by giving distinctive names to each full moon throughout the year. These names were also applied to the entire month in which they occurred. There are considerable variations in such names across the Northern Hemisphere, but here is a selection of the most popular ones:

**Legendary Names of Full Moons**

| | |
|---|---|
| January | Winter Moon, Wolf Moon |
| February | Snow Moon, Hunger Moon |
| March | Lantern Moon, Crow Moon |
| April | Egg Moon, Planter's Moon |
| May | Flower Moon, Milk Moon |
| June | Rose Moon, Strawberry Moon |
| July | Thunder Moon, Hay Moon |
| August | Grain Moon, Green Corn Moon |

| September | Harvest Moon, Fruit Moon |
| October | Hunter's Moon, Falling Leaves Moon |
| November | Frosty Moon, Freezing Moon |
| December | Christmas Moon, Long Night Moon |

*Fomalhaut.* During late evenings in September, the first-magnitude star Fomalhaut may be seen low down in the south. It is easy to locate, since two of the stars in the Great Square of Pegasus (Beta and Alpha) point downwards to it (Figure 19). Fomalhaut is the only prominent

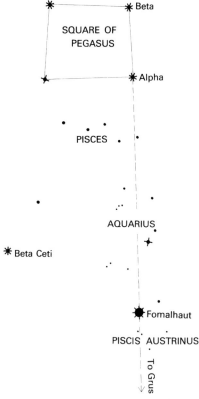

**Figure 19.** The bright star Fomalhaut in the constellation of Piscis Austrinus, the Southern Fish, is the most southerly first-magnitude star that can be seen from the United Kingdom. It may be found relatively easily by extending a line downwards from Beta and Alpha Pegasi (the two right-hand stars of the Great Square), as shown here.

star in the little group of Piscis Austrinus, the Southern Fish (also known as Piscis Australis). Indeed, the name Fomalhaut means the 'fish's mouth'.

Because Fomalhaut is so far south, it is never seen to advantage in Britain, and from northern Scotland it scarcely rises at all. On the other hand, it is extremely conspicuous as seen from the southern parts of the United States. Not far from it is the constellation of Grus, the Crane, which has a distinctive outline and several rather bright stars, but which is to all intents and purposes invisible from Europe.

Fomalhaut is one of the Sun's nearer neighbours, lying at a distance of twenty-five light years. This is almost the same distance as the brilliant Vega in Lyra, a similar bluish-white star of spectral type A. But Fomalhaut is a lower-mass star than Vega, with a smaller diameter and lower surface temperature, and so Fomalhaut has a luminosity only sixteen times that of our Sun, compared with more than fifty times for Vega. Consequently, Fomalhaut appears more than a magnitude fainter than Vega, although from Britain it appears even fainter on account of its low altitude; Vega passes almost overhead on summer evenings.

In 1983, the Infra-Red Astronomy Satellite (IRAS) discovered that there was far more infrared radiation coming from Fomalhaut than expected. This radiation appears to be coming from a huge disk of cool dust and gas around the star, very similar to that also found around Vega, and possibly a forming planetary system.

# October

*New Moon:* 3 October          *Full Moon:* 17 October

*Summer Time* in the United Kingdom ends on 30 October.

MERCURY is not suitably placed for observation by those in the latitudes of the British Isles. Further south it becomes available for observation in the evenings after the first few days of the month. For observers in southern latitudes this will be the most favourable evening apparition of the year. Figure 20 shows, for observers in latitude 35°S, the changes in azimuth (true bearing from the north through east, south and west) and altitude of Mercury on successive evenings when the Sun is 6° below the horizon. This condition is known as the end of evening civil twilight and in this latitude and at this time of year occurs about thirty minutes after sunset. The changes in the brightness of the planet are indicated by the relative sizes of the circles marking Mercury's position at five-day intervals. It will be noticed that Mercury is at its brightest before it reaches greatest eastern elongation early next month. Its magnitude remains near zero throughout October.

VENUS, magnitude −4.3, continues to be visible as a brilliant object low in the south-western sky in the early evenings.

MARS, its magnitude brightening from −1.7 to −2.3 during the month, continues to be visible as a very conspicuous object in the evening sky as soon as it is dark, by the end of October. The planet reaches its first stationary point on 1 October, and for the rest of the month moves retrograde from Taurus back into Aries. During this apparition the closest approach of the planet to the Earth occurs on 31 October, when it is 69 million kilometres (43 million miles) distant, a full week before it comes to opposition.

JUPITER, magnitude −1.7, may possibly be seen by observers in the Southern Hemisphere low above the south-western horizon shortly

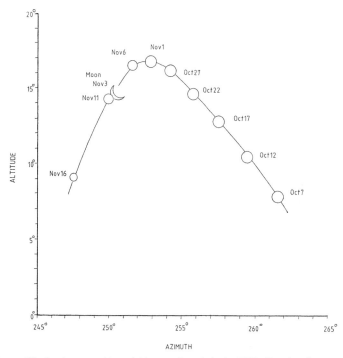

**Figure 20.** Evening apparition of Mercury, from latitude 35°S. (Angular diameters of Mercury and the crescent Moon not to scale.)

after sunset for a brief while, but only during the first week of the month. Jupiter passes through conjunction on 22 October and therefore remains unsuitably placed for observation during the rest of the month.

**SATURN,** continues to be visible in the eastern sky in the early mornings. The planet has a magnitude of +0.3 and therefore is noticeably brighter than Castor and Pollux which have magnitudes of +1.9 and +1.1 respectively.

*Vulcanoids?* Mercury is well placed in the evening sky for observers in southern latitudes during the second half of October, and Venus is also visible after sunset. Venus is spectacular, but Mercury is much less conspicuous because it is always near the Sun – so near, in fact, that it can never be surveyed with the Hubble Space Telescope, and we rely mainly upon the results from the one spacecraft that has passed and imaged it (Mariner 10, in March and September 1974 and March 1975). A new space mission, MESSENGER, is now on its way, but is not due to make its first fly-by of Mercury until January 2008, and as yet we have maps of less than half of the surface.

It was once believed that there was a planet closer in than Mercury moving at around 21 million kilometres from the Sun in a period of 19 days 17 hours. This was based upon the observation that Mercury's elliptical path around the Sun shifts slightly with each orbit such that its point of closest approach to the Sun (perihelion) precesses over time. All planetary orbits precess in this way, but the amount of the precession for all the planets *except* Mercury could be predicted precisely using Newton's equations governing planetary motion. Urbain Le Verrier, the French astronomer whose calculations had led to the discovery of Neptune in 1846, believed that an unseen planet was responsible for the discrepancy, and it was even given a name: Vulcan, after the Roman god of fire and metallurgy. However, it was never identified, and the precession of Mercury's orbit was eventually explained by Einstein's General Theory of Relativity, whose predictions exactly matched the observations.

Vulcan is a myth. However, many comets pass close in to the Sun, and there are also some asteroids whose orbits take them into these torrid regions. But can there be asteroids that remain close in? They may exist, and are referred to as Vulcanoids.

Obviously, such small bodies will be exceedingly hard to find if they exist, because they orbit so near the Sun. Ground-based searches for Vulcanoids have been carried out during total solar eclipses, and for the brief twilight period just before sunrise and after sunset when any Vulcanoids would be above the horizon, but the Sun is not. However, the relative faintness of Vulcanoids against the twilight sky, coupled with atmospheric unsteadiness and haze, hamper ground-based searches and no Vulcanoids have been discovered.

Nevertheless, a modest number – perhaps a few hundred – relatively small rocky objects, from 1 to 20 kilometres across, could have survived from the very earliest times to the present day, far inside the orbit of Mercury. Lying so close to the Sun, Vulcanoids could contain traces of the first materials to form within the inner Solar System, and would be very important for studies of the origin of the planets.

It would be very interesting to find them, and in 2003 a determined search was made using a most unexpected vehicle – a two-seater F/A-18B Hornet jet fighter aircraft flying through the stratosphere at over Mach 0.9. From an altitude of 15 kilometres (49,000 feet) over the Mojave Desert, the view of the twilight sky near the Sun was far darker and clearer than could be obtained from the ground

Using a sensitive camera from the back seat of the fighter, Daniel Burda and Alan Stern of the Southwest Research Institute at Boulder, Colorado, planned to make several flights and to do their best to track down any Vulcanoids in the bright near-Sun sky. Observations are conducted with a modified miniature CCD video camera, fitted with a high-quality, fast, 85-mm zoom lens. The camera, called the Southwest Ultraviolet Imaging System–Aircraft (SWUIS–A), originally conceived for use on the Space Shuttle, shoots image-intensified frames at video rates of sixty frames a second, and sends the data to an onboard video recorder. The imaging process eliminates any movement of the image due to the motion of the aircraft itself, and also allows the researchers to obtain what is tantamount to a time exposure. So far, no Vulcanoids have been found, but there is no theoretical bar to their existence and the hunt will continue.

It is also worth noting that Mercury and Venus, unlike all the other principal planets of the Solar System, have no satellites. Careful searches for a Mercurian satellite have been fruitless, but there was an 'alarm' on 27 March 1974, two days before the first Mariner 10 fly-by. An instrument on the probe began to record bright emissions in the extreme ultraviolet region; they vanished, but then reappeared, and a satellite was suspected. Alas, it was found that the object was an ordinary star, 31 Crateris. If Mercury had a satellite even a few kilometres across it would surely have been found by now.

The case of the phantom satellite of Venus is different and, as with Vulcan, a name was allotted: Neith. The object was first reported in 1666 by G.D. Cassini who was probably the best planetary observer of the time (Cassini discovered four of Saturn's satellites as well as the main division in the rings). Other reports followed up to 1874. However, it is certain that the observers were deceived by telescopic 'ghosts'. Phobos and Deimos, the tiny attendants of Mars, must surely be captured asteroids rather than bona fide satellites.

*This Month's Eclipse of the Sun.* The second solar eclipse of 2005 is on 3 October. It will be an annular eclipse, where the Moon appears smaller than the Sun as it passes centrally across the solar disc. A bright ring, or 'annulus', of sunlight – the so-called 'ring of fire' – will be seen at greatest eclipse. The eclipse path first reaches land near the border of northern Portugal and Spain, and crosses the Iberian Peninsula, passing right over Spain's capital city Madrid. It then traverses the Mediterranean Sea (crossing the southernmost part of the island of Ibiza) making landfall again in Algeria, and tracks south-eastwards though Tunisia, Libya, the north-eastern corner of Chad, the Sudan, the south-western corner of Ethiopia, Kenya and Somalia, ending in the Indian Ocean south of the Seychelles. Greatest eclipse with 4m 31s of annularity occurs in the Sudan. Residents of Madrid, Spain will see an annular eclipse lasting 4m 11s, and in Algiers the duration of the annular phase will be 3m 47s. A partial eclipse will be seen across much of Europe, including the British Isles. From southern England, including London, almost 60 per cent of the solar disk will be covered at maximum eclipse on the morning of 3 October.

# November

MERCURY reaches greatest eastern elongation (24°) on 3 November and three weeks later passes through inferior conjunction. It is not visible to observers as far north as the British Isles, but further south continues to be available for observation in the early evening until the middle of the month. During this period its magnitude fades from −0.2 to +1.1. Observers should see the diagram given with the notes for October.

VENUS has been slowly increasing in brightness during the past few months and is now at magnitude −4.5. It reaches its maximum eastern elongation (47°) on 3 November, and is visible in the south-western sky after sunset. Although still best seen from southern latitudes, visibility conditions for observers in the latitudes of the British Isles are improving and by the end of the month Venus can be seen for nearly two hours after sunset.

MARS, magnitude −2.3, is visible throughout the hours of darkness as it comes to opposition on 7 November. Observers will note that for a few weeks it is actually brighter than Jupiter. Mars is moving retrograde in the constellation of Aries.

JUPITER, magnitude −1.7, is too close to the Sun for observation at first but, after the first week of the month for those in the Northern Hemisphere, and after the second week of the month for those in the Southern Hemisphere, becomes visible low above the south-eastern horizon for a short while before sunrise. Jupiter is in the constellation of Virgo, moving slowly eastwards, a few degrees east of Spica.

SATURN is still visible in the eastern and southern skies, rising a little north of east several hours before midnight for observers in the latitudes of the British Isles, but near midnight itself for Southern

Hemisphere observers. It is moving very slowly direct in the constellation of Cancer but reaches its first stationary point on 22 November, when it starts to retrograde. Its magnitude is +0.2.

*Voyager 1 at Saturn.* Twenty-five years ago this month, in November 1980, the Voyager 1 spacecraft made its remarkable and highly successful fly-by of the planet Saturn – an encounter described by the *Sunday Telegraph* in London as 'the most spectacular piece of space exploration since man first stepped on the Moon'.

Voyager 1 had passed Jupiter in March 1979, taking advantage of that giant planet's gravity to speed it on towards Saturn, and by the summer of 1980 it was already close enough to its target to begin acquiring its first images of the planet. The trajectory of Voyager 1 had been carefully chosen so that it would make a close encounter with Saturn's largest moon, Titan, considered almost as important an objective as Saturn itself.

As Voyager 1 approached Saturn in October 1980, the rings were revealed to have an unexpected amount of fine detail, and the cloud structure of Saturn began to appear. On 6 October came the first major surprise – a new kind of feature in Saturn's bright B ring: dark bands that extended radially from the planet like the spokes of a wheel. Later that month, while making a movie of these dark spokes, scientists were to discover two new moons, orbiting on the inner and outer edges of the extremely narrow F ring. These were the first two new satellites of Saturn to be discovered by Voyager. By early November, the steadily improving images revealed more and more narrow belts in Saturn's atmosphere, and a small red spot in the southern hemisphere, similar to, but much smaller than, Jupiter's Great Red Spot. Voyager 1 also obtained a tantalizing far view of Saturn's enigmatic satellite Iapetus, which has a leading hemisphere that is much darker than the trailing one. Another small moon was also found, just beyond the outer edge of the A ring, and the Cassini Division no longer appeared empty, but filled with a variety of light and dark rings.

Titan was to be the first major encounter of the Voyager 1 fly-by, on 11 November 1980, the day before the closest approach to Saturn itself. To everyone's surprise, the orange clouds enveloping Titan were so dense and opaque that there was no sign of the surface beneath. Later measurements showed the surface temperature to be around −180 degrees Celsius, with a surface pressure of about 1.6 bars. The primary constituent of Titan's atmosphere was revealed to be nitrogen, in addition to the hydrocarbons and photochemical smog, with layers of haze above the major cloud deck. As scientists remarked at the time, it would only be possible to learn about Titan's surface by using an orbiting imaging radar or a probe that could parachute through the clouds and make a landing. (By early 2005, that prophecy should hopefully have become a reality with the radar mapping of Titan's surface by Cassini and the landing on the surface by the ESA's Huygens probe. For more information, see the article by Garry E. Hunt elsewhere in this *Yearbook*.)

Once past Titan, Voyager 1's trajectory took it past many of Saturn's lesser satellites: Tethys on the way in towards the closest approach to Saturn, and Mimas, Enceladus, Dione, Rhea and Hyperion on the way out. All of the densities appeared low, suggesting that these satellites are

**Figure 21.** On 12 November 1980, eight hours after closest approach to Saturn, Voyager 1 acquired this image of the planet's ring system. From left to right: the narrow, bright F-ring, the A-ring, the Cassini Division, the broad B-ring, and the C-ring (dark grey area). The unique lighting brings out the many hundreds of bright and dark ringlets in the rings. The dark spoke-like features discovered by Voyager 1 during its approach to Saturn now appear as bright streaks in the B-ring. (Image courtesy of NASA Jet Propulsion Laboratory.)

**Figure 22.** The crescent of Saturn, the planet's rings and their shadows are seen in this Voyager 1 image acquired on 13 November 1980 at a distance of 1.5 million kilometres as the spacecraft began to leave the Saturn system. The bright limb of Saturn is clearly visible through the A, B, and C rings. The dark band cutting through the crescent is the shadow of the rings. (Image courtesy of NASA Jet Propulsion Laboratory.)

basically icy bodies with only small quantities of rocky materials. Many surface details were resolved on the minor satellites – there were craters aplenty – and there were surprises, such as the huge crater on Mimas and the highly irregular shape of Hyperion. Parts of Enceladus are rather smooth and appear to have been resurfaced, with large craters obliterated, possibly by some kind of 'cryovolcanism' (the icy equivalent of what we call volcanic activity).

During the close encounter phase, Voyager 1 also made many important observations of Saturn's rings (Figure 21), from both above and below the ring plane, revealing their intricate structure and quite unexpected complexity: thousands of individual small ringlets and narrow gaps. On leaving Saturn behind (Figure 22), Voyager 1 was moving on a trajectory that would take it far beyond the Solar System, and out into interstellar space. As far as the planets were concerned, Voyager 1's job was done, but its close-range observations of Titan had been of crucial importance. Their success meant that Voyager 1's sister craft, Voyager 2, could now be kept on a path that would take it through the Saturnian system and onwards to survey the remaining two giant planets: Uranus in January 1986 and Neptune in August 1989. In paving the way for the next phase of this great voyage of exploration, the importance of Voyager 1's success at Saturn cannot be overstated.

*The Mystery of Achernar.* For observers in southern latitudes, the bright star Achernar is well placed, high in the southern sky during November evenings. Achernar is the leading star of the long and winding constellation of Eridanus, the River, which begins close to Rigel in Orion, the Hunter, and meanders its way down to the far southern sky, ending at Achernar (Figure 23). Though it is one of the largest constellations in the entire sky, Eridanus contains remarkably few bright stars. Apart from Achernar, there are only three stars above magnitude 3.0: Beta, or Kursa (magnitude 2.8), Theta, or Acamar (2.9) and Gamma, or Zaurak (3.0). Mythologically, Eridanus is linked with the River Po – this was the river into which the young Phaethon plunged when he lost control of the Sun-chariot he was driving for the day, and Jupiter was forced to strike him down with a thunderbolt.

Achernar, with a declination of 57°S, does not rise from anywhere north of latitude 33°N, so it is invisible from the whole of Europe and most of the United States, except for Florida and southern Texas (where it only just rises) and Hawaii. From Cairo, Achernar merely skirts the southern horizon when at its highest. However, it is circumpolar from Buenos Aires, Cape Town and Sydney, and from all of New Zealand. Achernar is the ninth brightest star in the sky at magnitude 0.46 – only very slightly fainter than Procyon in the pattern of the Little Dog. Achernar is of spectral type B5 and so is blue-white in colour; it is 400 times as luminous as the Sun, and about eighty-five light years distant.

There is a minor mystery connected with Achernar. The name Achernar comes from the Arabic words meaning 'the end of the river', but originally was this Achernar at all? Ptolemy does not mention it, and the name seems to have been applied to a different star, Theta Eridani (now called Acamar), which could be easily seen from Alexandria where Ptolemy lived. Nobody is quite sure which was the

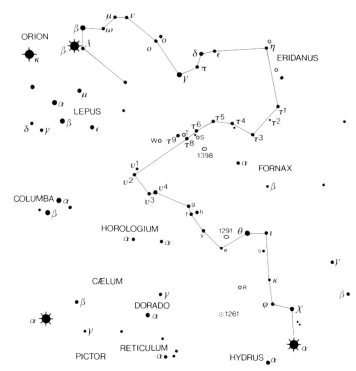

**Figure 23.** The meandering constellation of Eridanus, the River, begins near the brilliant Rigel in Orion, and ends with the star Achernar, which lies at the extreme southern end of the pattern. Theta Eridani (Acamar), which is now of the third magnitude, was rated as first magnitude by Ptolemy, but there may have been confusion with Achernar, and any real fading seems unlikely.

original Achernar, and to make matters worse there is some suggestion that Theta has declined in brightness from the first magnitude since ancient times – which does seem rather unlikely. Incidentally, Theta Eridani is a fine double star; both components are white, of magnitudes 3.4 and 4.5, with a separation of 8.3 arcseconds.

# December

*New Moon:* 1 & 31 December   *Full Moon:* 15 December

*Solstice:* 21 December

**MERCURY** is a morning object, though for observers in the latitudes of the British Isles it is lost to view for the last week of the month. For observers in northern temperate latitudes this will be the most favourable morning apparition of the year. Figure 24 shows, for observers in latitude 52°N, the changes in azimuth (true bearing from the north through east, south and west) and altitude of Mercury on successive mornings when the Sun is 6° below the horizon. This condition is known as the beginning of morning civil twilight and in this latitude and at this time of year occurs about thirty-five minutes before sunrise. The changes in the brightness of the planet are indicated by

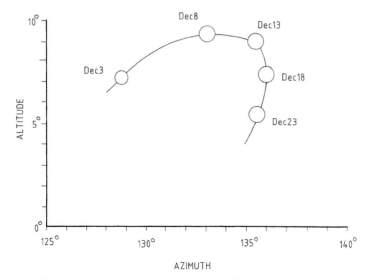

**Figure 24.** Morning apparition of Mercury, from latitude 52°N.

the relative sizes of the circles marking Mercury's position at five-day intervals. It will be noticed that Mercury is at its brightest after it reaches greatest western elongation (21°) on 12 December. During the month its magnitude brightens from +1.2 to −0.5.

VENUS attains its greatest brilliancy, magnitude −4.7, on 9 December, and as it is still some 40° from the Sun it dominates the south-western sky in the early evenings. Observers in the latitudes of the British Isles will find it is visible for about one and a half to two hours after sunset. Observers with telescopes will note that the crescent shape becomes thinner and thinner during the month, but at the same time the apparent diameter almost doubles in size.

MARS, just past opposition, is still visible for the greater part of the hours of darkness. It may be seen in the southern skies as soon as the evening twilight has faded sufficiently and even at the end of the year will be visible well after midnight. Mars is in Aries, reaching its second stationary point on 10 December, and then resuming its direct motion. During the month its brightness decreases noticeably from a magnitude of −1.6 to −0.6.

JUPITER is a brilliant morning object, magnitude −1.8. It is visible in the south-eastern quadrant of the sky for several hours before sunrise. Early in December Jupiter moves from Virgo into Libra.

SATURN, magnitude 0.0, is approaching opposition at the end of next month and so is visible to observers in the Northern Hemisphere from the late evening onwards. Those in the Southern Hemisphere will have to wait until after midnight before it becomes suitably placed for observation. Saturn is in the constellation of Cancer.

*Old Ideas About Saturn's Rings.* With Saturn now visible in the late evening sky, and with its rings still well displayed, this may be a suitable moment to say something about very early observations of them. Galileo observed the rings in 1610, but did not realize their nature; he believed Saturn to be a triple planet, and was very surprised when, in 1612, he found that he could no longer see the companions. This was in fact hardly surprising, as the rings were edge-on during that year. Later, Galileo re-observed them, but he never solved the mystery.

Observers during the first part of the seventeenth century were no more successful. Hevelius of Danzig believed Saturn to be elliptical in shape, with two appendages attached to the surface. The French mathematician de Roberval suggested that Saturn must be surrounded by a hot zone giving off vapours, reflecting sunlight off the edges but appearing opaque when seen in depth. Christopher Wren – at this stage a professional astronomer, with fame as an architect yet to come – thought that Saturn had an elliptical corona, meeting the globe in two places, and was rotating with Saturn once in each sidereal period. However, before Wren published his theory, the Dutch astronomer Christiaan Huygens (who also discovered Saturn's largest satellite, Titan, in 1655) correctly stated that the cause of the strange appearance was 'a flat ring'. Characteristically, Wren accepted this at once. Others did not. Honoré Fabri, a French Jesuit, claimed that the aspect of Saturn could be explained by the movements of four satellites – two dark and two bright. A pamphlet published by Divini, though probably written by Fabri, was little more than a vitriolic attack on Huygens. The Fabri theory was easily refuted, and within a few years all astronomers accepted the ring hypothesis. Even Fabri was forced to do so in 1665.

*The Winter Solstice.* This year the northern winter solstice falls on 21 December at 18.35 Universal Time. The dates of the solstices are not fixed, but are defined as the time when the Sun, in its apparent yearly path around the sky (the ecliptic), reaches its greatest declination north or south of the celestial equator. On 21 December, the Sun will be at its furthest south of the celestial equator, and will be overhead at local noon for places lying on the Tropic of Capricorn. Of course, for people living in the Southern Hemisphere, winter and summer solstices are reversed, so that for them 21 December marks the summer solstice.

In northerly latitudes, the winter solstice marks the first day of the season of winter. The winter solstice is the shortest day of the year, in the sense that the length of time between sunrise and sunset on this day is a minimum for the year. The midday Sun is at its lowest, and from now on will start to gain altitude. During late December, then, the northern part of the globe has less daylight and more darkness than at any other time of the year. After the winter solstice, those people who live in the Arctic, such as parts of northern Canada and northern Scandinavia, can begin to look forward to the return of sunlight.

# Eclipses in 2005

During 2005 there will be four eclipses, two of the Sun and two of the Moon.

1. *An annular-total eclipse of the Sun on 8 April* is visible as a partial eclipse from New Zealand's North Island, the Pacific Ocean, Central and most of South America, southern North America, the Caribbean and Bermuda. The partial phase begins at 17ʰ 51ᵐ and ends at 23ʰ 20ᵐ. Near the beginning of the eclipse's central path, the eclipse is annular. For most of the track over the southern Pacific Ocean the eclipse is total, but it again becomes annular before crossing Panama and northern Colombia. The annular eclipse ends in Venezuela. Annularity begins at 18ʰ 53ᵐ and ends at 22ʰ 18ᵐ. The maximum duration of totality (42 seconds) occurs in the middle of the Pacific Ocean.

2. *A penumbral eclipse of the Moon on 24 April* visible only from the western parts of North America, New Zealand and eastern Australia.

3. *An annular eclipse of the Sun on 3 October* is visible as a partial eclipse from Greenland, Iceland, the Arctic Ocean, the eastern North Atlantic Ocean, Africa (except the southern tip), Madagascar, Europe (including the British Isles), west and south Asia and the Indian Ocean. The partial phase begins at 07ʰ 35ᵐ and ends at 13ʰ 28ᵐ. The path of annularity commences in the eastern North Atlantic Ocean, crosses northern Portugal, Spain, northern Algeria, Tunisia, Libya, north-east Chad, Sudan, south-west Ethiopia, northern Kenya and the extreme south of Somalia before finishing in the Indian Ocean. Annularity begins at 08ʰ 41ᵐ and ends at 12ʰ 23ᵐ. The maximum duration of annularity is 4ᵐ 31ˢ, from the Sudan.

4. *A minor partial eclipse of the Moon on 17 October* is visible from the Arctic Ocean, North and Central America, western South America, the Pacific Ocean, Australasia, the Indian Ocean, Asia and the Southern Ocean. The partial eclipse begins at 11ʰ 34ᵐ and ends at 12ʰ 32ᵐ. Even at maximum eclipse, only 7 per cent of the lunar surface is obscured.

# Occultations in 2005

In the course of its journey round the sky each month, the Moon passes in front of all the stars in its path, and the timing of these occultations is useful in fixing the position and motion of the Moon. The Moon's orbit is tilted at more than 5° to the ecliptic, but it is not fixed in space. It twists steadily westwards at a rate of about 20° a year, a complete revolution taking 18.6 years, during which time all the stars that lie within about 6.5° of the ecliptic will be occulted. The occultations of any one star continue month after month until the Moon's path has twisted away from the star, but only a few of these occultations will be visible from any one place during the hours of darkness.

There are fifteen occultations of bright planets in 2005, one of Mercury, three of Venus, two of Mars, and nine of Jupiter.

Only four first-magnitude stars are near enough to the ecliptic to be occulted by the Moon: these are Aldebaran, Regulus, Spica and Antares. Both Spica and Antares undergo occultation in 2005.

Predictions of these occultations are made on a worldwide basis for all stars down to magnitude 7.5, and sometimes even fainter. The British Astronomical Association has produced a complete lunar occultation-prediction package for personal-computer users. Occultations of stars by planets (including minor planets) and satellites have aroused considerable attention.

The exact timing of such events gives valuable information about positions, sizes, orbits, atmospheres and sometimes of the presence of satellites. The discovery of the rings of Uranus in 1977 was the unexpected result of the observations made of a predicted occultation of a faint star by Uranus. The duration of an occultation by a satellite or minor planet is quite small (usually of the order of a minute or less). If observations are made from a number of stations it is possible to deduce the size of the planet.

The observations need to be made either photoelectrically or visually. The high accuracy of the method can readily be appreciated when one realizes that even a stopwatch timing accurate to a tenth of a second is, on average, equivalent to an accuracy of about 1 kilometre in the chord measured across the minor planet.

# Comets in 2005

The appearance of a bright comet is a rare event that can never be predicted in advance, because this class of object travels round the Sun in enormous orbits with periods that may well be many thousands of years. There are therefore no records of the previous appearances of these bodies, and we are unable to follow their wanderings through space.

Comets of short period, on the other hand, return at regular intervals, and attract a good deal of attention from astronomers. Unfortunately they are all faint objects, and are recovered and followed by photographic methods using large telescopes. Most of these short-period comets travel in orbits of small inclination that reach out to the orbit of Jupiter, and it is this planet that is mainly responsible for the severe perturbations that many of these comets undergo. Unlike the planets, comets may be seen in any part of the sky, but since their distances from the Earth are similar to those of the planets their apparent movements in the sky are also somewhat similar, and some of them may be followed for long periods of time.

The ephemerides of three comets that should be visible to observers with small telescopes in 2005 are given below:

## Comet Tempel 1

| Date 2005 | 2000.0 RA | | Dec. | | Distance from Earth | Distance from Sun | Elong-ation from Sun | Mag. |
|---|---|---|---|---|---|---|---|---|
|  | h | m | ° | ′ | AU | AU | ° |  |
| May 10 | 12 | 53.1 | + 9 | 43 | 0.714 | 1.607 | 137.0 | + 9.8 |
| 20 | 12 | 52.0 | + 7 | 04 | 0.728 | 1.576 | 129.1 | + 9.6 |
| 30 | 12 | 55.2 | + 3 | 52 | 0.751 | 1.550 | 122.1 | + 9.5 |
| Jun 9 | 13 | 02.6 | + 0 | 17 | 0.783 | 1.529 | 115.9 | + 9.5 |
| 19 | 13 | 13.9 | − 3 | 34 | 0.822 | 1.515 | 110.6 | + 9.5 |
| 29 | 13 | 28.8 | − 7 | 30 | 0.868 | 1.508 | 106.0 | + 9.5 |
| Jul 9 | 13 | 46.9 | −11 | 27 | 0.920 | 1.507 | 102.0 | + 9.7 |
| 19 | 14 | 07.8 | −15 | 16 | 0.979 | 1.512 | 98.6 | + 9.8 |

**Comet 2003 T4 (LINEAR)**

| Date 2005 | | RA h m | Dec. ° ′ | | Distance from Earth AU | Distance from Sun AU | Elong-ation from Sun ° | Mag. |
|---|---|---|---|---|---|---|---|---|
| Jan | 0 | 18 08.6 | +39 | 23 | 2.026 | 1.806 | 63.0 | +10.1 |
| | 10 | 18 34.9 | +35 | 08 | 1.965 | 1.674 | 58.4 | + 9.7 |
| | 20 | 18 59.4 | +30 | 53 | 1.905 | 1.542 | 53.7 | + 9.3 |
| | 30 | 19 22.4 | +26 | 36 | 1.842 | 1.412 | 49.2 | + 8.8 |
| Feb | 9 | 19 44.4 | +22 | 12 | 1.772 | 1.285 | 45.2 | + 8.3 |
| | 19 | 20 05.8 | +17 | 32 | 1.689 | 1.163 | 42.1 | + 7.8 |
| Mar | 1 | 20 27.6 | +12 | 23 | 1.592 | 1.052 | 40.2 | + 7.2 |
| | 11 | 20 51.0 | + 6 | 26 | 1.479 | 0.958 | 39.8 | + 6.7 |
| | 21 | 21 18.3 | − 0 | 39 | 1.357 | 0.888 | 40.8 | + 6.1 |
| | 31 | 21 52.6 | − 9 | 06 | 1.235 | 0.853 | 43.3 | + 5.8 |
| Apr | 10 | 22 38.5 | −18 | 45 | 1.135 | 0.857 | 46.8 | + 5.6 |
| | 20 | 23 40.3 | −28 | 25 | 1.082 | 0.900 | 51.0 | + 5.7 |
| | 30 | 0 57.1 | −35 | 54 | 1.096 | 0.976 | 55.1 | + 6.1 |
| May | 10 | 2 18.2 | −39 | 32 | 1.178 | 1.075 | 58.3 | + 6.7 |
| | 20 | 3 29.3 | −39 | 53 | 1.310 | 1.189 | 60.0 | + 7.3 |
| | 30 | 4 24.9 | −38 | 37 | 1.473 | 1.312 | 60.4 | + 8.0 |
| Jun | 9 | 5 07.0 | −36 | 57 | 1.651 | 1.440 | 59.9 | + 8.7 |
| | 19 | 5 39.4 | −35 | 28 | 1.832 | 1.570 | 59.0 | + 9.3 |
| | 29 | 6 05.1 | −34 | 22 | 2.008 | 1.702 | 58.0 | + 9.8 |

**Comet 2003 K4 (LINEAR)**

| Date 2005 | | RA h m | Dec. ° ′ | | Distance from Earth AU | Distance from Sun AU | Elong-ation from Sun ° | Mag. |
|---|---|---|---|---|---|---|---|---|
| Jan | 10 | 5 03.7 | −52 | 08 | 1.287 | 1.745 | 99.7 | + 6.5 |
| | 20 | 4 14.2 | −43 | 04 | 1.469 | 1.864 | 97.0 | + 7.0 |
| | 30 | 3 49.6 | −35 | 05 | 1.699 | 1.984 | 91.4 | + 7.6 |
| Feb | 9 | 3 37.6 | −28 | 39 | 1.957 | 2.104 | 84.5 | + 8.2 |
| | 19 | 3 32.4 | −23 | 33 | 2.228 | 2.224 | 77.0 | + 8.7 |
| Mar | 1 | 3 31.2 | −19 | 30 | 2.501 | 2.344 | 69.5 | + 9.2 |

**Comet 2003 K4 (LINEAR)** – *cont'd*

| Date 2005 | RA | | Dec. | | Distance from Earth | Distance from Sun | Elong- ation from Sun | Mag. |
|---|---|---|---|---|---|---|---|---|
| | h | m | ° | ′ | AU | AU | ° | |
| Mar 11 | 3 | 32.5 | −16 | 13 | 2.769 | 2.463 | 62.0 | + 9.6 |
| 21 | 3 | 35.4 | −13 | 32 | 3.026 | 2.581 | 54.7 | +10.0 |

The following comets are expected to return to perihelion in 2005, and to be brighter than magnitude +15.

| Comet | Year of Discovery | Period (years) | Predicted Date of Perihelion 2005 |
|---|---|---|---|
| 10/P Tempel 2 | 1873 | 5.4 | Feb 15 |
| 2003 T4 (LINEAR) | 2003 | ? | Apr 3 |
| 1983 V1 P/Hartley–IRAS) | 1983 | 21.5 | Jun 25 |
| 21/P Giacobini–Zinner | 1900 | 6.6 | Jul 2 |
| 9/P Tempel 1 | 1867 | 5.5 | Jul 5 |
| 37/P Forbes | 1929 | 6.3 | Aug 1 |
| 101/P Chernykh | 1977 | 13.9 | Dec 24 |

# Minor Planets in 2005

Although many thousands of minor planets (asteroids) are known to exist, only a few thousand of them have well-determined orbits and are listed in the catalogues. Most of these orbits lie entirely between the orbits of Mars and Jupiter. All these bodies are quite small, and even the largest, Ceres, is only 913 kilometres (567 miles) in diameter. Thus, they are necessarily faint objects, and although a number of them are within the reach of a small telescope few of them ever attain any considerable brightness. The first four that were discovered are named Ceres, Pallas, Juno and Vesta. Actually the largest four minor planets are Ceres, Pallas, Vesta and Hygeia. Vesta can occasionally be seen with the naked eye, and this is most likely to happen when an opposition occurs near June, since Vesta would then be at perihelion. Below are ephemerides for Ceres, Pallas, Juno and Vesta in 2005.

## 1 Ceres

| | | | 2000.0 RA | | Dec. | | Geo-centric Distance | Helio-centric Distance | Phase Angle | Visual Magni-tude | Elong-ation |
|---|---|---|---|---|---|---|---|---|---|---|---|
| | | h | m | ° | ′ | | | | ° | | ° |
| Jan | 30 | 15 | 09.94 | − 8 | 25.6 | | 2.551 | 2.617 | 21.9 | 8.6 | 82.8W |
| Feb | 9 | 15 | 19.98 | − 8 | 53.4 | | 2.425 | 2.623 | 22.1 | 8.5 | 90.3W |
| | 19 | 15 | 28.33 | − 9 | 12.3 | | 2.300 | 2.630 | 21.8 | 8.3 | 98.2W |
| Mar | 01 | 15 | 34.71 | − 9 | 22.7 | | 2.178 | 2.637 | 21.1 | 8.2 | 106.5W |
| | 11 | 15 | 38.80 | − 9 | 25.2 | | 2.061 | 2.644 | 19.8 | 8.1 | 115.4W |
| | 21 | 15 | 40.35 | − 9 | 21.1 | | 1.954 | 2.651 | 18.0 | 7.9 | 124.8W |
| | 31 | 15 | 39.21 | − 9 | 11.8 | | 1.859 | 2.659 | 15.5 | 7.7 | 134.7W |
| Apr | 10 | 15 | 35.37 | − 8 | 59.2 | | 1.782 | 2.666 | 12.4 | 7.5 | 145.1W |
| | 20 | 15 | 29.09 | − 8 | 45.9 | | 1.726 | 2.674 | 8.9 | 7.3 | 155.8W |
| | 30 | 15 | 20.96 | − 8 | 34.7 | | 1.694 | 2.682 | 5.2 | 7.1 | 165.9W |
| May | 10 | 15 | 11.82 | − 8 | 28.7 | | 1.688 | 2.690 | 3.4 | 7.0 | 170.9W |
| | 20 | 15 | 02.71 | − 8 | 30.8 | | 1.710 | 2.698 | 5.8 | 7.1 | 164.3E |
| | 30 | 14 | 54.62 | − 8 | 42.8 | | 1.757 | 2.706 | 9.4 | 7.4 | 154.1E |
| Jun | 9 | 14 | 48.29 | − 9 | 05.5 | | 1.829 | 2.714 | 12.8 | 7.6 | 143.7E |

**1 Ceres** – *cont'd*

| | | RA | | Dec. | | Geo-centric Distance | Helio-centric Distance | Phase Angle | Visual Magni-tude | Elong-ation |
|---|---|---|---|---|---|---|---|---|---|---|
| | | h | m | ° | ′ | | | ° | | ° |
| Jun | 19 | 14 | 44.23 | − 9 | 39.1 | 1.920 | 2.722 | 15.7 | 7.8 | 133.6E |
| | 29 | 14 | 42.64 | −10 | 22.3 | 2.028 | 2.730 | 18.0 | 8.0 | 124.0E |
| Jul | 9 | 14 | 43.47 | −11 | 13.8 | 2.149 | 2.739 | 19.7 | 8.2 | 115.0E |
| | 19 | 14 | 46.61 | −12 | 12.1 | 2.279 | 2.747 | 20.8 | 8.4 | 106.6E |
| | 29 | 14 | 51.82 | −13 | 15.4 | 2.415 | 2.755 | 21.4 | 8.5 | 98.6E |
| Aug | 8 | 14 | 58.89 | −14 | 22.2 | 2.554 | 2.764 | 21.5 | 8.7 | 91.0E |
| | 18 | 15 | 07.61 | −15 | 31.0 | 2.694 | 2.772 | 21.3 | 8.8 | 83.7E |

**2 Pallas**

| | | RA | | Dec. | | Geo-centric Distance | Helio-centric Distance | Phase Angle | Visual Magni-tude | Elong-ation |
|---|---|---|---|---|---|---|---|---|---|---|
| | | h | m | ° | ′ | | | ° | | ° |
| Jan | 00 | 12 | 17.09 | −11 | 25.4 | 1.964 | 2.213 | 26.4 | 8.6 | 91.1W |
| | 10 | 12 | 27.22 | −10 | 46.5 | 1.848 | 2.228 | 25.8 | 8.4 | 99.2W |
| | 20 | 12 | 35.25 | − 9 | 38.6 | 1.735 | 2.243 | 24.7 | 8.3 | 108.0W |
| | 30 | 12 | 40.88 | − 7 | 57.3 | 1.629 | 2.260 | 22.7 | 8.1 | 117.6W |
| Feb | 9 | 12 | 43.82 | − 5 | 39.4 | 1.534 | 2.278 | 20.0 | 7.9 | 128.0W |
| | 19 | 12 | 43.90 | − 2 | 44.1 | 1.456 | 2.297 | 16.3 | 7.7 | 139.2W |
| Mar | 1 | 12 | 41.22 | + 0 | 43.9 | 1.400 | 2.316 | 12.0 | 7.4 | 150.9W |
| | 11 | 12 | 36.18 | + 4 | 33.3 | 1.371 | 2.337 | 7.5 | 7.2 | 162.2W |
| | 21 | 12 | 29.55 | + 8 | 26.5 | 1.372 | 2.358 | 4.7 | 7.1 | 168.9W |
| | 31 | 12 | 22.40 | +12 | 03.6 | 1.405 | 2.379 | 6.9 | 7.3 | 163.4E |
| Apr | 10 | 12 | 15.83 | +15 | 08.5 | 1.466 | 2.401 | 11.0 | 7.6 | 152.8E |
| | 20 | 12 | 10.80 | +17 | 31.9 | 1.553 | 2.424 | 14.8 | 7.9 | 141.8E |
| | 30 | 12 | 07.91 | +19 | 12.5 | 1.661 | 2.447 | 18.0 | 8.2 | 131.4E |
| May | 10 | 12 | 07.42 | +20 | 14.1 | 1.784 | 2.470 | 20.3 | 8.4 | 121.8E |
| | 20 | 12 | 09.33 | +20 | 42.3 | 1.918 | 2.494 | 21.9 | 8.6 | 113.0E |
| | 30 | 12 | 13.43 | +20 | 43.6 | 2.060 | 2.518 | 22.9 | 8.9 | 104.8E |

**3 Juno**

| | | RA | | Dec. | | Geo-centric Distance | Helio-centric Distance | Phase Angle | Visual Magni-tude | Elong-ation |
|---|---|---|---|---|---|---|---|---|---|---|
| | | h | m | ° | ′ | | | ° | | ° |
| Sep | 17 | 4 | 57.50 | + 9 | 06.0 | 1.556 | 1.990 | 29.9 | 8.8 | 99.7W |
| | 27 | 5 | 10.32 | + 7 | 49.7 | 1.458 | 1.985 | 29.0 | 8.7 | 106.1W |
| Oct | 7 | 5 | 20.67 | + 6 | 20.7 | 1.365 | 1.981 | 27.7 | 8.5 | 112.9W |
| | 17 | 5 | 28.13 | + 4 | 42.4 | 1.281 | 1.979 | 25.8 | 8.3 | 120.2W |
| | 27 | 5 | 32.32 | + 2 | 59.6 | 1.206 | 1.979 | 23.3 | 8.1 | 127.9W |
| Nov | 6 | 5 | 32.98 | + 1 | 19.4 | 1.145 | 1.981 | 20.3 | 7.9 | 136.0W |
| | 16 | 5 | 30.17 | − 0 | 09.4 | 1.098 | 1.985 | 17.0 | 7.7 | 144.0W |
| | 26 | 5 | 24.40 | − 1 | 17.0 | 1.071 | 1.991 | 14.0 | 7.6 | 150.8W |
| Dec | 6 | 5 | 16.63 | − 1 | 54.5 | 1.063 | 1.999 | 12.1 | 7.5 | 154.8W |
| | 16 | 5 | 08.32 | − 1 | 56.2 | 1.077 | 2.009 | 12.4 | 7.6 | 154.0E |
| | 26 | 5 | 00.96 | − 1 | 22.5 | 1.113 | 2.020 | 14.6 | 7.7 | 148.9E |

**4 Vesta**

| | | RA | | Dec. | | Geo-centric Distance | Helio-centric Distance | Phase Angle | Visual Magni-tude | Elong-ation |
|---|---|---|---|---|---|---|---|---|---|---|
| | | h | m | ° | ′ | | | ° | | ° |
| Jul | 9 | 5 | 00.04 | +19 | 13.5 | 3.383 | 2.568 | 11.8 | 8.5 | 31.2W |
| | 19 | 5 | 17.32 | +19 | 40.9 | 3.313 | 2.569 | 13.6 | 8.5 | 36.6W |
| | 29 | 5 | 34.29 | +20 | 00.9 | 3.233 | 2.570 | 15.3 | 8.5 | 42.1W |
| Aug | 8 | 5 | 50.87 | +20 | 13.6 | 3.142 | 2.571 | 17.0 | 8.5 | 47.7W |
| | 18 | 6 | 06.92 | +20 | 19.9 | 3.041 | 2.572 | 18.5 | 8.4 | 53.5W |
| | 28 | 6 | 22.31 | +20 | 20.4 | 2.932 | 2.572 | 19.8 | 8.4 | 59.5W |
| Sep | 7 | 6 | 36.91 | +20 | 16.2 | 2.815 | 2.571 | 20.9 | 8.3 | 65.8W |
| | 17 | 6 | 50.52 | +20 | 08.5 | 2.691 | 2.570 | 21.9 | 8.2 | 72.3W |
| | 27 | 7 | 02.96 | +19 | 58.8 | 2.562 | 2.569 | 22.5 | 8.2 | 79.1W |
| Oct | 7 | 7 | 14.01 | +19 | 48.7 | 2.429 | 2.568 | 22.9 | 8.1 | 86.4W |
| | 17 | 7 | 23.39 | +19 | 40.2 | 2.295 | 2.566 | 22.8 | 7.9 | 94.0W |
| | 27 | 7 | 30.82 | +19 | 35.6 | 2.162 | 2.563 | 22.3 | 7.8 | 102.2W |
| Nov | 6 | 7 | 35.95 | +19 | 36.9 | 2.032 | 2.561 | 21.2 | 7.6 | 111.1W |
| | 16 | 7 | 38.45 | +19 | 46.4 | 1.909 | 2.558 | 19.5 | 7.4 | 120.5W |
| | 26 | 7 | 38.02 | +20 | 05.5 | 1.798 | 2.554 | 17.0 | 7.2 | 130.8W |
| Dec | 6 | 7 | 34.48 | +20 | 35.0 | 1.702 | 2.550 | 13.8 | 7.0 | 141.8W |
| | 16 | 7 | 27.91 | +21 | 13.9 | 1.626 | 2.546 | 9.9 | 6.8 | 153.6W |
| | 26 | 7 | 18.78 | +21 | 59.1 | 1.576 | 2.541 | 5.4 | 6.6 | 166.0W |

# Meteors in 2005

Meteors ('shooting stars') may be seen on any clear moonless night, but on certain nights of the year their number increases noticeably. This occurs when the Earth chances to intersect a concentration of meteoric dust moving in an orbit around the Sun. If the dust is well spread out in space, the resulting shower of meteors may last for several days. The word 'shower' must not be misinterpreted – only on very rare occasions have the meteors been so numerous as to resemble snowflakes falling.

If the meteor tracks are marked on a star map and traced backwards, a number of them will be found to intersect in a point (or a small area of the sky) that marks the radiant of the shower. This gives the direction from which the meteors have come.

The following table gives some of the more easily observed showers with their radiants; interference by moonlight is shown by the letter M.

| Limiting Dates | Shower | Maximum | RA | | Dec. | |
|---|---|---|---|---|---|---|
| | | | h | m | ° | |
| Jan 1–6 | Quadrantids | Jan 3 | 15 | 28 | +50 | M |
| April 19–25 | Lyrids | Apr 22 | 18 | 08 | +32 | M |
| May 1–8 | Aquarids | May 4 | 22 | 20 | −01 | |
| June 17–26 | Ophiuchids | June 19 | 17 | 20 | −20 | M |
| July 15–Aug 15 | Delta Aquarids | July 29 | 22 | 36 | −17 | M |
| July 15–Aug 20 | Piscis Australids | July 31 | 22 | 40 | −30 | |
| July 15–Aug 20 | Capricornids | Aug 2 | 20 | 36 | −10 | |
| July 23–Aug 20 | Perseids | Aug 12 | 3 | 04 | +58 | |
| Oct 16–27 | Orionids | Oct 20 | 6 | 24 | +15 | M |
| Oct 20–Nov 30 | Taurids | Nov 3 | 3 | 44 | +14 | |
| Nov 15–20 | Leonids | Nov 17 | 10 | 08 | +22 | M |
| Dec 7–16 | Geminids | Dec 13 | 7 | 32 | +33 | M |
| Dec 17–25 | Ursids | Dec. 22 | 14 | 28 | +78 | M |

# Some Events in 2006

## ECLIPSES

There will be four eclipses, two of the Sun and two of the Moon.

| | |
|---|---|
| 14 March: | Penumbral eclipse of the Moon |
| 29 March: | Total eclipse of the Sun – Africa, Europe, Asia |
| 7 September: | Partial eclipse of the Moon – Australasia, Asia, Africa, Europe |
| 22 September: | Annular eclipse of the Sun – Central and South America, Africa |

## THE PLANETS

*Mercury* may be seen more easily from northern latitudes in the evenings about the time of greatest eastern elongation (24 February) and in the mornings about the time of greatest western elongation (25 November). In the Southern Hemisphere the corresponding most favourable dates are around 8 April (mornings) and 17 October (evenings).

*Venus* is visible in the evenings in early January. From mid January to October it is visible in the mornings. From mid December until the end of the year it is visible in the evenings.

*Mars* does not come to opposition in 2006.

*Jupiter* is at opposition on 4 May in Libra.

*Saturn* is at opposition on 27 January in Cancer.

*Uranus* is at opposition on 5 September in Aquarius.

*Neptune* is at opposition on 11 August in Capricornus.

*Pluto* is at opposition on 16 June in Serpens.

## TRANSIT

There will be a transit of *Mercury* on 8 November.

# Part II

# Article Section

# Mars: The New Wave of Exploration

PETER J. CATTERMOLE

The year 2004 has proved to be one of the most exciting ever for devotees of the Red Planet. Not since the second half of 1997, when NASA's rover Sojourner trundled across the plains at the mouth of Ares Vallis, and Mars Global Surveyor orbited the planet above it, has there been so much activity on Mars. For the first time in history, two mobile rovers are exploring the surface of another planet at the same time: Spirit started making wheel tracks halfway around Mars from Terra Meridiani on 15 January 2004, while Opportunity had done so at Meridiani a few days earlier. 'We're two for two! One dozen wheels on the soil,' JPL's Chris Lewicki, flight director, announced to the control room in Pasadena, California. Add to this the three orbiting spacecraft – NASA's Mars Odyssey and Mars Global Surveyor and ESA's Mars Express – and you have a potent laboratory of tools for studying the Red Planet.

## EXCITING RESULTS FROM THE THREE ORBITERS

Mars Global Surveyor, oldest of the currently active spacecraft, has been incredibly successful. It has sent back a huge number of images and continues to obtain data of the highest significance; it has been used recently to assist mission controllers in tracking down where later spacecraft have ended up on the surface and also to select suitable landing sites for the future. It was Global Surveyor that discovered the quite unexpected grey haematite layer in the southern highlands of the planet, detected magnetic striping in the Martian Southern Hemisphere similar to that developed on the Earth's ocean floors, and sent back high-resolution images of landforms clearly etched by running water.

Mars Odyssey, a more sophisticated spacecraft, launched by NASA during 2001, has joined Global Surveyor and has already been monitoring Mars for over one Martian year. Odyssey carries a number of multispectral devices that enable scientists to chart seasonal changes on the planet, in particular the retreat and advance of the polar ice caps and the amounts of water and carbon dioxide, both in the Martian atmosphere and on its surface. Furthermore, the high-resolution camera is transmitting incredibly detailed images of Mars, of special significance being a magnificent series of images of the planet's south polar region.

One of the most interesting new pieces of information collected by Mars Global Surveyor's Thermal Emission Spectrometer (TES) concerns the iron–magnesium silicate, olivine. This mineral, which weathers easily in the presence of water, has been found in abundance in certain parts of Mars, and has been mapped in detail, on a global scale, by scientists at the United States Geological Survey (USGS). The presence of olivine implies that chemical weathering by water is at a low level on the planet and that Mars may have been cold and dry throughout most of its geologic history. In addition, USGS scientists have identified details in the infrared spectrum that indicate what is probably a sulphate in the pervasive dust and many of the bright soils on Mars.

This recent geochemical information has been supplemented by information from Mars Odyssey's THEMIS instrument. The first overview analysis of a year's worth of high-resolution infrared data gathered by its THermal EMission Imaging System (THEMIS) was published in the journal *Science* in June 2003. THEMIS is providing planetary geologists with detailed temperature and infrared radiation images of the Martian surface. The images reveal geological details that were impossible to detect even with the high-resolution Mars Orbital Camera on Mars Global Surveyor and have 300 times higher resolution than MGS's Thermal Emission Spectrometer.

By analysing the spectra from the ten different bands of infrared light the instrument can detect, the THEMIS team has begun to identify specific mineral deposits, including a significant layer of the mineral olivine near the bottom of a canyon four and a half kilometres deep known as Ganges Chasma near the eastern end of the vast Valles Marineris trough system. The olivine occurs in darker basaltic volcanic rocks that cover a large portion of the planet. The sulphates occur in brighter rocks that are most likely mechanically weathered and with trace amounts of fine-grained haematite. The fine-grained haematite probably also comes from mechanical grinding of coarser-grained volcanic rocks. The origin of the coarse-grained hematite is still unknown.

Although there is as yet no consensus, it appears likely that abundant water probably exists below the surface, but only in a frozen state and rarely, if ever, has it existed at the surface in a liquid form (certainly not in recent times). This makes searching for life much more difficult. However, the European Space Agency's (ESA) Mars Express spacecraft, launched in June 2003, has onboard radar instrumentation that is able to probe a few kilometres beneath the surface, and should provide, for the first time, information about how much and in what form $H_2O$ exists at these depths.

On arriving at Mars, Mars Express released the first ever European lander, Beagle 2, on Christmas Eve 2003. Although Beagle was

**Figure 1.** A meandering channel, Reull Vallis, once formed by flowing water on Mars. This image was obtained by the High Resolution Stereo Camera (HRSC) onboard ESA's Mars Express orbiter on 15 January 2004 from a height of 273 kilometres. (Image courtesy of ESA/DLR/FU and G. Neukum.)

successfully released, contact was regrettably lost shortly afterwards and it is still unclear what happened. Most likely it failed to brake sufficiently and impacted the surface. However, Mars Express continues to orbit Mars and is sending back exciting data from its high-resolution stereo camera (HRSC), which operates in full colour and 3D. Furthermore, OMEGA, a visible and infrared mineralogical mapping spectrometer, is determining mineralogical composition in 100-metre squares, and SPICAM, an ultraviolet and infrared atmospheric spectrometer, is analysing atmospheric composition, in particular ozone and water vapour. Additionally, the spacecraft carries a planetary Fourier spectrometer (PFS), energetic neutral atoms analyser (ASPERA) and radio science experiment (MaRS), plus a subsurface sounding radar/altimeter (MARSIS). The latter is particularly interesting since, as mentioned above, it can penetrate to a depth of a few kilometres, thereby allowing scientists to build up a picture of any layering in the Martian subsurface and possible water ice.

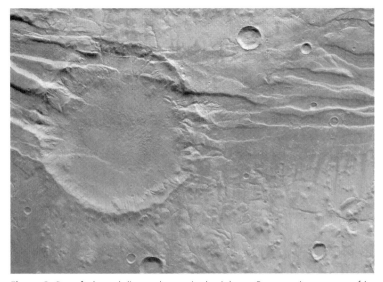

**Figure 2.** Deep faults and disrupted crater in the Acheron Fossae region, an area of intensive tectonic activity in the past. This image was obtained by the HRSC on Mars Express. The large crater is about 55 kilometres in diameter. The rifting crosses the older impact crater with at least three alternating horsts and grabens. (Image courtesy of ESA/DLR/FU and G. Neukum.)

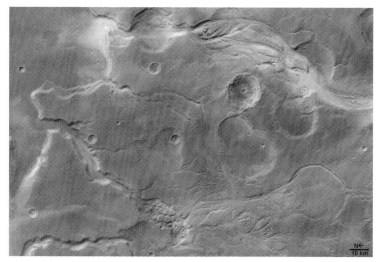

**Figure 3.** Evidence of flash-flooding at Mangala Valles on Mars. This image of fluvial surface features at Mangala Valles was obtained by the HRSC on board ESA's Mars Express spacecraft. (Image courtesy of ESA/DLR/FU and G. Neukum.)

One of its most exciting results was the discovery of the spectral fingerprint of water ice, for the first time. Through the initial mapping of the south polar cap on 18 January 2004, OMEGA, the combined camera and infrared spectrometer, revealed the presence of both water and carbon dioxide ices. This confirmed similar results collected by earlier NASA missions. The first PFS data also show that the carbon dioxide distribution is different between the Northern and Southern Hemispheres of Mars.

## 'SPIRIT' AND 'OPPORTUNITY' START WORK

The last of the new Mars arrivals were NASA's two powerful new-generation Mars Exploration Rovers, Spirit and Opportunity. The spacecraft carrying these were launched from Cape Canaveral on 10 June and 7 July 2003 respectively, and both have dispatched their payloads immaculately. Spirit and Opportunity are larger, more mobile and better equipped than the 1997 Mars Pathfinder rover; each carries a sophisticated set of instruments to search for evidence that might decide whether past environments at selected sites were wet enough to be hospitable to life. The two rovers are identical, but landed on different regions of Mars: Spirit came down in Gusev Crater, located along the boundary of the planetary dichotomy (the line separating the northern plains and southern plateau), while Opportunity landed on Meridiani Planum. Engineering data from both Spirit and Opportunity are returned in relays via NASA's Mars Odyssey orbiter.

Thus, during January 2004, three orbiting and two roving vehicles were at work, trying to decipher some of the remaining mysteries of Mars. Hopefully they will continue to send back information for many months and, if we combine their results, we may be able to generate new ideas about the planet and how it has evolved. However, if the past is anything to on, the more data that is returned from Mars, the more questions are asked.

The lander vehicle of the Mars Exploration Rover (MER) mission features a design dramatically different from Mars Pathfinder's, for whereas Pathfinder had scientific instruments on both the lander and the small Sojourner rover, Spirit and Opportunity carried all of their instruments with them. Soon after each landing, the lander began making a reconnaissance of its landing site by taking a 360° visible colour and infrared image panorama. Unfortunately, problems were encountered with Spirit soon after it had done this, but round-the-clock work at Mission Control resulted in a solution to the rover's computer problems and eventually it left its base and moved down on to the surface. Opportunity also encountered some problems with its heater, but is now operating well and continues to flourish.

The first 360° colour panorama delivered by Spirit revealed some rather fine-grained fragmental material close to the rover (the Martian equivalent of 'sand'), several large rocks, a small impact crater about 200 metres in diameter and some nearby low hills. Opportunity returned the first pictures of its landing site about four hours after reaching Mars. The pictures indicated that the spacecraft sat in a shallow crater about twenty metres across. Mission Control was particularly excited about this lucky landing, as it immediately put the instruments in a position to study bedrocks thrown out from the impact that formed this crater many millions of years ago. Initial images revealed very fine-grained red sediment, many small rocks and some slabby outcrops close by, together with the interior wall of the crater itself.

While the flight team needed only seven days following Opportunity's landing to get the rover off its lander, it required twelve days for Spirit earlier in the same month, since Spirit's deflated airbags had been an obstacle to the craft rolling away from its lander probe. So, in late January 2004, Opportunity drove on to the soil of Meridiani Planum. New science results from the rover had already indicated that the site did indeed have crystalline haematite – the principal reason the site was selected for exploration – and cheers erupted at Mission Control when Opportunity sent a picture looking back at the now-empty lander and showing wheel tracks in the very fine-grained red Martian regolith.

Grey granules covering most of the crater floor surrounding Opportunity were shown by the rover's miniature thermal emission spectrometer to contain haematite; this infrared-sensing instrument is used for identifying rock types from a distance. The same material was also detected at the Spirit site. Crystalline haematite is of special interest because, on Earth, it often forms under wet environmental conditions. The concentration of haematite appeared to be strongest in a layer of dark material above a light-coloured outcrop in the wall of the crater where Opportunity sits. It should be noted, however, that not all haematite owes its presence to an origin in bodies of water, since much terrestrial iron oxide can be deposited from volcanic lavas and juvenile waters percolating into the crust from great depth. Interestingly, clay was not detected at the site; thus, although the loose sediment is of very fine grain size (typical of terrestrial clay minerals), it does not appear to have been generated by alteration and hydration of volcanic rocks or impact breccias.

## OBJECTIVES FOR THE ROVERS

During the three-month primary mission during which both rovers were active – their life was governed by their solar-powered batteries – rocks and soils were analysed with a suite of five instruments, while a special tool called the Rock Abrasion Tool, or 'RAT', was used to expose fresh rock surfaces for study. Using images and measurements received daily from the rovers, scientists command the vehicle to go to rock and soil targets of interest and evaluate their composition and their texture at microscopic scales. Initial targets were near the landing sites, but later targets were much further afield. Each Mars Exploration

Rover was expected to travel much further than the distance achieved by Mars Pathfinder during its three-month prime mission, and new records for roving across the Martian surface were indeed set. Each Mars Exploration Rover has a mass of nearly 180 kilograms and was expected to travel up to about 40 metres per sol, or Martian day. In the event, the two rovers exceeded all expectations, and by May 2004 they had established one-sol driving records of over 90 metres for Spirit and 140 metres for Opportunity.

One of the original objectives of the MER mission was to determine the history of climate and water at two sites on Mars where conditions may once have been favourable to life. The two landing sites were selected carefully on the basis of intensive study of orbital data collected by earlier orbiting missions. Both sites show strong evidence for the presence of past water. Spirit landed in Gusev Crater, a wide-impact basin that may once have held a lake; Opportunity's destination was Meridiani Planum, an area about halfway around Mars from Gusev that has a broad outcrop of an iron oxide – grey haematite – that may form in the presence of water. The rovers' instruments are currently studying the geological and geochemical features at both sites with a view to evaluating whether conditions could have been suitable for the development of life.

**Figure 4.** This pair of pieced-together images was taken by the Spirit rover's rear navigation camera on 6 March 2004. It reveals the long and rocky path of nearly 240 metres that Spirit had travelled since safely arriving at Gusev Crater on 3 January 2004. The lander can still be seen in the distance, but will never be 'home' again for the journeying rover. (Image courtesy of NASA Jet Propulsion Laboratory.)

**Figure 5.** The rim and interior of the crater nicknamed 'Bonneville' dominate this 180° mosaic of images taken by the panoramic camera of the Spirit rover on 12 March 2004. (Image courtesy of NASA Jet Propulsion Laboratory and Cornell University.)

Gusev, the site for Spirit, is an impact crater 150 kilometres in diameter located along the boundary between the northern plains and southern uplands. Into it, from the south, runs the major flood channel, Ma'adim Vallis, which at its maximum is twenty-five kilometres wide and two kilometres deep. Landforms on the interior floor of the crater suggest, quite strongly, the deposition of stratified rocks and, bearing in mind the connection between Gusev and the outflow channel, the flow from this channel may have become ponded inside the crater, forming a huge lake. Close study of such rocks would, firstly, establish whether this is so or not, and also enable scientists to study the nature of rocks brought down from the highlands in floods.

A small impact crater on Mars became the new home for NASA's Opportunity rover, and a larger crater lies nearby. 'If it got any better, I couldn't stand it,' said Dr Doug Ming, rover science team member from NASA's Johnson Space Center, Houston, as the news of the rover's exact landing spot broke. With the instruments on the rover and just the rocks and soil within the small crater, it was hoped that Opportunity would enable scientists to determine which of several theories about the region's past environment was right, he said. Scientists value such crater locations as a way to see what is beneath the surface without needing to dig. Scientists had hoped for a specific landing site where they could examine both the surface layer that is rich in haematite and an underlying geological feature of light-coloured layered rock. The small crater appears to have exposures of both, with soil that could be the haematite unit and an exposed outcropping of the lighter rock layer, and the scientists are considering whether the haematite may have formed in a long-lasting lake or in a volcanic environment.

An even bigger crater, which could provide access to deeper layers for more clues to the past, lies nearby. Images taken by a camera on the bottom of the lander during Opportunity's final descent showed that a crater about 150 metres across was likely to be within about one kilometre or half a mile of the landing site. Scientist Steve Squyres, at one of the early press conferences in January 2004, presented an outline for Opportunity's potential activities. After driving off the lander, the rover would first examine the soil adjacent to the lander, then drive to the outcrop of layered-looking rocks and spend considerable time examining it. Then the rover would climb out of the small crater, take a look around, and head for the bigger crater. If it achieves all this,

**Figure 6.** Layered rocks near the Opportunity rover's landing site at Meridiani Planum, on Mars. Visible on two of the rocks are holes drilled by the rover. These enabled scientists to study the underlying chemical composition, providing clues to a water-soaked past. Opportunity's wheel tracks can be seen at the bottom left and right sides of this image. (Image courtesy of NASA Jet Propulsion Laboratory.)

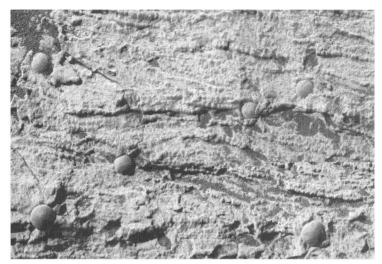

**Figure 7.** Magnified view from the Opportunity rover of a portion of a Martian rock showing fine layers (laminae) that are truncated, discordant and at angles to each other. These indicate that the sediments which formed the rock were most likely laid down in flowing water. (Image courtesy of NASA Jet Propulsion Laboratory, Cornell University and US Geological Survey.)

our knowledge of this particular region of Mars will undoubtedly be greatly enhanced but, more importantly, vital new information about Mars's global geology will have been obtained. The new data, plus that obtained from Spirit, within the Gusev Crater, coupled with the wealth of new and high-resolution information recorded by ESA's Mars Express orbiter, Mars Global Surveyor and Mars Odyssey, should mean that 2004 turns out to be a very good year for Mars. Whether it will be good for those in search of life remains to be seen.

# Astronomy and Art

PAUL MURDIN

## THE WORK OF VAN GOGH

When the artist Vincent van Gogh travelled as a young man from the grey skies of his native Holland to the sunnier lands of France, first to Paris and then to the clear skies of Provence, he became more and more aware of the colours in the countryside around him and, in those completely light-unpolluted days, of the colours in the stars. In a letter of 19 June 1888, he wrote:

> In the blue depth the stars were sparkling greenish, yellow, white, pink, more brilliant, more sparkling gem-like than at home – even in Paris: opals you might call them, emeralds, lapis lazuli, rubies, sapphires.
>
> Certain stars are citron-yellow, others have a pink glow, or a green, blue and forget-me-not brilliance. And without my expatiating on this theme it will be clear enough that putting little white dots on a blue-black surface is not enough.

His interests in astronomy were developed as a result of his growing thoughts about the relationship of man and the cosmos, and he began to paint pictures in which the cosmos figured, in part at least to put the human affairs portrayed in the same picture into a broader perspective. The greatest example was *Starry Night*, which is in the Museum of Modern Art in New York. The picture shows a French hamlet under a night sky with a crescent moon, Venus (a bright star) and a swirling Milky Way. The village is dwarfed by the majesty of the sky above and even the church spire, representing man's greatest spiritual achievements, is puny.

The painting can be dated from letters that Vincent sent to his sister and brother, and must have been completed about 16–18 June 1889 in Saint-Rémy. According to Charles Whitney, Venus was a morning star at the time and the Milky Way would have been seen in the east during the evening, rising high before dawn. Vincent mentioned the 'blue whiteness' of the Milky Way in a letter to his brother Theo the previous summer, so he was familiar with the Milky Way. The dates of the crescent moon do not, however, tally with the dates of the painting. On those dates the Moon was gibbous, between full and three quarters. The Moon is painted in the corner of the picture, obviously positioned and shaped for reasons of composition.

Vincent took similar liberties with the positions of celestial objects in *Road with Cypress and Star* in the Kroller-Muller State Museum in Otterlo in the Netherlands. It shows a thin crescent moon and two bright stars in a more or less horizontal row. From letters, art historians have concluded that the picture was painted shortly before Vincent left Saint-Rémy on 16 May 1890. Donald Olson has found that on 20 April of that year the waxing crescent moon was near Venus and Mercury on that date, in the same orientation – but reversed left to right. The planetary grouping was spectacular enough to have been forecast in the press as a celestial sight. Vincent presumably reversed the orientation for compositional reasons. He did the same thing in an earlier picture painted in the autumn of 1888, called *Starry Night on the Rhône* in the Musée d'Orsay in Paris. It shows the Great Bear at night over the Rhone river. The landscape is viewed to the south-east, where the Great Bear never appears, because it is a northern circumpolar constellation. Vincent moved the stars to make a better picture.

## GREAT COMETS IN ART

Sometimes, the dating of the astronomical phenomena in a painting is very easy. The full title of William Dyce's picture *Pegwell Bay*, in the Tate Gallery in London, is *Pegwell Bay: A Recollection of 5th October 1858*. The painting shows low tide under the chalk cliffs of this seaside resort in Kent, where Dyce and his family holidayed in the autumn of 1858, and in the sky above hangs Donati's Comet C/1858 L1. The painting is an allegory of time, representing the rhythm of the tides, the geological aeons over which the chalk strata were laid down and the vast reaches of astronomical time, punctuated by the precise arrival of the Comet. The Comet was a grand sight, one of the most beautiful ever seen and the first to be photographed, by the Harvard University

astronomer George Bond and, the night before, by an English commer-
cial photographer called William Usherwood of Surrey, both of them
recording a small smudge on the collodion plates in use at the time. In
Dyce's painting, over the rock pools are bent Victorian ladies, gentle-
men and children, gathering seashells, paying insufficient attention
to the larger wonders of nature around them. I think scientists may do
this, too. They concentrate on the side details of some study and should
make time to see how it fits into the larger picture.

Possibly the greatest comet of the nineteenth century was the Great
September Comet, C/1882 R1. It was discovered on 1 September 1882
by a group of Italian sailors in the Southern Hemisphere. Over the
next few days several astronomers in Australia, New Zealand, South
Africa and South America independently rediscovered it, but such
was the difficulty of communication from these distant countries that
it was not notified to European observatories until 12 September. By
14 September it was visible in daylight at noon. At the time of its passage
through perihelion on 17 September it was discovered in England by
Ainslee Common. It was described by Sir David Gill at the Royal Observatory at the Cape as 'of astonishing brilliancy as it rose behind the
mountains to the east of Table Bay and seemed in no way to diminish
in brightness when the Sun rose a few minutes afterwards'. Gill used an
ordinary portrait camera to photograph the comet. The photograph
revealed not only the comet but also the images of great numbers of
stars in the background, and Gill realized that the stars could be better
mapped *en masse* by photography than by individually sighting on
each one. This led to the Carte du Ciel project to record the whole sky.

The Great September Comet was visible in the dawn sky from
Britain during gaps in the stormy weather of the time. One of the artists
who drew it was Edward Poynter. He saw it from Scotland and drew it,
with its magnificent curved and hollow-looking tail, on 23 October
1882. At the time he was working on a composition depicting the death
of Julius Caesar. The picture, the *Ides of March,* which is in the City Art
Gallery in Manchester, was shown at an exhibition at the Royal
Academy in the spring of 1883. Caesar and his wife Calpurnia stand
dramatically within a Roman colonnade in the imperial palace, looking
out into the dawn sky above the Roman skyline. Calpurnia is clutching
Caesar's toga and gesturing fearfully to a comet in the sky. The picture
illustrates Shakespeare's play, *Julius Caesar,* when Calpurnia warns
Caesar with the words:

When beggars die there are no comets seen;
The heavens themselves blaze forth the death of princes.

Shakespeare in fact got this a bit wrong. There *was* a comet in the sky around the time of Caesar's death on the Ides of March (15 March), the Comet of 44 BC. But the comet was in the sky just after Caesar's death, and was commonly supposed to indicate that after his assassination Caesar had been admitted to the company of the gods. Anyway, Poynter was working on the picture in various sketches in 1881 and 1882. As he was finishing the painting for exhibition, the appearance of the Great September Comet inspired him to paint it into the picture to stand in for the Comet of 44 BC.

## BETTER THAN REALISTIC

All these pictures show the astronomical universe in a way that is better than realistic – at least, better at making a picture. I thought at first when I began to look at such things that the artists had got it wrong in that they were not representing the real universe. I now realize that even in trying to depict the celestial bodies correctly astronomers depict things in a way that is better than real. Take the representation of the Moon. The Moon can be represented better than accurately. Approaching this aim from both sides of the arts–science divide were the portraitist John Russell (1745–1806) and the engineer James Nasmyth (1808–90).

In 1874, Nasmyth, a mechanical engineer who invented the steam hammer and the altazimuth telescope configuration that bears his name, constructed plaster of Paris models of the lunar surface that were photographed under oblique solar illumination (some of the models are in the Science Museum in London). The pictures, with long exposures due to the technology then available, were obtained under studio conditions, not through wobbly telescopes in motion and not looking through the atmosphere. The models are (of course!) smaller in scale than the craters and mountain ranges that they represent, and omit the progressive curvature of the Moon. This eliminates the natural change in angle of illumination that exists on the Moon, most oblique on the side of the feature near to the terminator and progressing towards overhead illumination on the side of the feature towards the centre of illumination of the Moon's face. Nasmyth's technique maintains the contrast of moonlit and shadowed areas of the features across the image. The models also exaggerate topographic contrast, like the more recent Magellan spacecraft images of the surface of the planet Venus (a five times exaggeration in height).

Nasmyth and Carpenter also prepared a map of the complete face of the Moon that used the same principle. Depicting the general aspect of each object, 'we so adjusted the shading that all objects should be shown under about the same angle of illumination – a condition that is never fulfilled upon the moon itself, but which we consider ourselves justified in exhibiting for the purpose of conveying a fair impression of how the various lunar objects actually appear at some one or other part of a lunation'.

Russell did something similar. From sketches made through Herschelian telescopes (in fact he was offered a telescope by William Herschel himself), he constructed pastel drawings of the gibbous moon 'painted from nature' but with impossibly oblique light over a larger region near the shadow's edge than happens in nature, in order to highlight the relief. There is one version of his picture in Soho House, Birmingham, associated with the members of the Lunar Society, and another in the Museum of the History of Science in Oxford.

These examples of the representation of the Moon have high astronomical content, which extends beyond the naturalistic and beyond the merely accurate. In more modern times, the Hubble Space Telescope pictures are photographic, yes, but realistic? I think not: even if you got to the Eagle Nebula in a space ship you would not find the Pillars of Creation looking like the Hubble picture (Figure 1). The colours of the original have been intensified, of course.

To return to the artist's job, it is to comment on why we do what we do. It is superstitious to fear the portents of the sky, like unheralded comets. We should go about our normal everyday business diligently and do our best to represent the scientific truth in some way that need not be merely representational. But we should also make time to be distracted from our small tasks to contemplate the grander scheme of things. In doing so, we shall undoubtedly realize the immensity of what we study and acknowledge our place in the universe, a small niche on a small planet in a big galaxy in a bigger universe. But the dialogue through art and science about all this is what makes us people big enough and important enough to face up to these realities in a human way.

**Figure 1.** These eerie, dark, pillar-like structures are columns of cool interstellar hydrogen gas and dust that are also incubators for new stars. They are part of the Eagle Nebula, a star-forming region 7,000 light years away in the constellation Serpens. The original colour image was constructed from three separate Hubble Space Telescope images, taken in the light of emission (red, green and blue) from different types of atoms. (Image courtesy of Jeff Hester and Paul Scowen (Arizona State University), and NASA/STScI.)

## GALLERY

You can visit these paintings not only in the Museums that I have mentioned, but also on the World Wide Web:

*Starry Night* is at moma.org/collection/depts/paint_sculpt/blowups/paint_sculpt_003.html.

*Starry Night over the Rhône* is at www.vangoghgallery.com/painting/p_0474.htm.

*Road with Cypress and Star* is at www.vangoghgallery.com/painting/p_0683.htm.

*Pegwell Bay* is at www.excelsiordirect.com/pegwell.htm.

*Ides of March* is at www.emory.edu/ENGLISH/classes/Shakespeare_Illustrated/Poynter.Ides.html.

Two of Russell's pastel drawings of the Moon are on display at the site of the Oxford Museum of the History of Science, but you have to use the search button for them (enter John Russell):

www.mhs.ox.ac.uk/images/index.htm.

Some of Nasmyth's photographs were exhibited in an exhibition which has a website (you need to look for the index to find Nasmyth's pictures of his models):

www.lhl.lib.mo.us/events_exhib/exhibit/ex_face_moon.shtml.

The Hubble Heritage picture of the Eagle Nebula is at:

hubblesite.org/gallery/showcase/nebulae/n6.shtml.

# Meteorites and Time: Aspects and Applications

JOE McCALL

Meteorites fall to Earth quite frequently, but only some five or six are recovered annually as rock or metal masses immediately after fall. Many are recovered by chance finds but cannot be connected to a known fall event. Some 30,000 meteorites are now recorded (Grady, 2000). The reason for the small number of recoveries from observed falls is that meteorites decay by natural weathering processes after they fall on to land, and such decay is very rapid in the temperate and well-vegetated terrains that cover much of the land surface of the globe. Obscuration by vegetation cover is also a bar to recovery.

The above statement concerns the *primary time relationship* of meteorites. However, there are many other time-related aspects of meteorites, some well known, others less so, and some surprising applications. I realized recently that the 'time relationships' of meteorites provides a framework for a comprehensive review. Below I present some of these aspects and applications.

## IDENTIFYING THE ASTEROID SOURCE

The time of fall to Earth of the 'bolide' accompanied by the fireball phenomenon is the basis of a novel project that was initiated late in 2003 in the Nullarbor Plain, Western Australia, a limestone desert terrain. The premise for this project is that in the vast majority of cases falls of meteorites are not well observed and so we cannot be sure of the 'whereabouts' in space the material comes from. To know this we need to match orbital characteristics with study of the meteorite material, but of the 30,000 meteorites now in collections, only a handful have associated orbits that are known. Meteorites were thought to come from comets until the Farmington fall, Kansas, 1890, was subjected to rigorous sifting of accounts of the event, which led later to deduction of asteroidal provenance (Zanda and Rotaru, 2001). Camera networks were also applied to this research – the 1959 Příbram fall in Czechoslovakia, for example, was orbitally related to an asteroidal source (Hutchison and Graham, 1994). Indeed, four of the six falls orbitally defined to date were recorded by camera networks.

The concept is a simple one: an array of well-spaced all-sky cameras (100–150 kilometres apart) records meteor tracks during the hours of darkness. Triangulation to fix the point of fall is possible if more than one camera records the event, and the size of the impactor can also be estimated. Yet, in four decades of such observation, only four actual recovered meteorite masses have in this way yielded orbital associations, among them the Lost City, USA, and Innisfree, Canada, falls. The reason? Vegetated and/or rough or unpopulated terrain makes the finding of the masses very difficult in temperate regions, where they decay rapidly, and no camera network has been mounted to date in an optimum area for recovery.

The Nullarbor Plain is one such optimum area. Several thousand meteorite masses have been recovered there. However, one single fall, at Camel Donga, produced 900 fusion-crust coated stones, and at Mulga North there were nearly as many, and these statistics show there must be numerous 'pairings' in this aggregate – masses that came in together. So a much lower number, some 300 separate arrivals, are represented by the above total, some 200 of them in Western Australia and some 100 in the extension of the Nullarbor Plain into South Australia (Figure 1). Such masses remain on the limestone surface with its thin, discontinuous clay mantle for long periods because rainfall is virtually nil, and the vegetation is very sparse (as the name implies). One iron meteorite, Mundrabilla, lay there for about a million years according to isotopic evidence (McCall, 1999), although the stony meteorites that form the bulk of the finds are shown by similar evidence to have been there for up to 36,000 years. The dark meteorites show up all over the plain in contrast against the whitish limestone. The area is very flat and cloud cover is thin and uncommon.

Scientists from Imperial College, London, working with Czech scientists, will firstly position a camera that has been developed to operate remotely for weeks without human intervention. Additional non-optical sensors will detect acoustic, infrasound and seismic effects

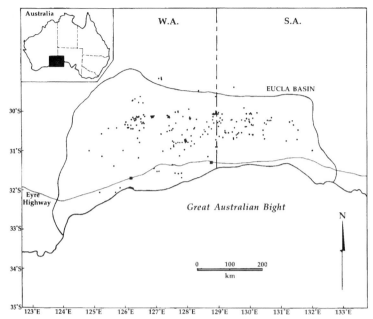

**Figure 1.** The Nullarbor Plain, showing where meteorite finds have been described and classified in Western and South Australia (from Bevan et al., 1998). Reprinted from *Quaternary Research*, 53, 2000, with permission from Elsevier.

of falls, which will allow 24-hour coverage. The first installation was at Kybo Station in the Central Nullarbor in October 2003. This extensive field test will, it is hoped, attract funds to complete the network.

It is estimated that four or five falls will be observed by such a network per year. Sites of falls can be fixed by the network to one to two square kilometres' accuracy. Between three and eight traversing workers will be required to search out the meteorite masses. Even if only a small proportion of the observed falls yield a meteorite mass, the team will be able to match the previous total of results obtained from camera networks within two or three years. Their results will be of immense scientific importance, for the aim is to correlate the orbits derived with those of known near-Earth asteroids. This would provide by proxy the equivalent of a sample return run to the asteroid at a small fraction of the cost of an actual robot space probe recovery mission to the asteroid.

## HISTORICAL TIME – SCIENTIFIC IMPORTANCE

Up to the end of the last decade of the eighteenth century and the first years of the nineteenth, the reality of 'showers of stones from the sky' was not accepted by science. This is surprising, because Diogenes of Apollonia came to the correct conclusion following a fall in 465 BC. He recognized the cosmic origin of meteorites and meteors (McCall, 1973): 'meteors are invisible stars that fall to Earth and die out, like the fiery stony star that fell to Earth near the Egos Potamos River'.

The earliest known observed fall, from which a meteorite mass is now preserved, was at Ensisheim, Alsace, in AD 1462, the mass being preserved in a church. The falls in Italy at Albareto in 1766 and Siena in 1794, though investigated by Troili and Sir William Hamilton, failed to overturn the lingering scepticism of science. As the colourful bishop the Earl of Bristol noted, the latter fall came the day after an eruption of Vesuvius, so it was easier to conclude that Vesuvius had hurled a shower of stones all the way to Siena than invoke cosmic agency (Fothergill 1974)! The Wold Cottage fall of a stone in Yorkshire in 1795, and analysis of both that and the Siena stone material by Howard, at last convinced scientists in England; and the L'Aigle rain of stones in 1803 convinced Biot in France. Nevertheless, the belief was still maintained that comets must be the source, and this belief was upheld until a later careful analysis of reported visual observations of the Farmington, Kansas, fall in 1890, photographic observations of the remarkably bright visual effects accompanying a fall at Příbram in Czechoslovakia in 1959, and later such evidence from three other falls indicated an asteroidal origin for the meteors.

Meteoritics had only a store of some two thousand actual specimens until the 1960s, when rabbit trappers on the Nullarbor Plain, a vast limestone desert in Western Australia, brought in numerous specimens: John Carlisle and his family discovered about eighty meteorites before his death a few years ago. While unofficial curator of the collections at the Western Australian Museum in 1967, I wrote prophetically: 'the Nullarbor Plain must be littered with meteorites' (McCall, 1967).

Shortly after this, to add to this new-found source of meteoritic wealth, a Japanese expedition found nine meteorites on an area of bare ice in Antarctica in 1969. This area, measuring only 5 × 10 kilometres, eventually yielded a thousand meteorites. Given that all things pertaining to space were 'in' at that time, these two events led to the development of systematic searches over desert regions – both hot deserts such as Nullarbor, North Africa and Arabia, and cold deserts, the blue ice developments in Antarctica. The result: the present total of 30,000 meteorites, representing about 20,000 arrival events.

The reality of *meteorite cratering* on Earth was even slower to be appreciated by science. Robert Hooke in 1665 considered the lunar craters and made some fine drawings of them, but dismissed them as volcanic, though he admitted the possibility of impacts by some foreign bodies. The matter rested until the 1890s when G.K. Gilbert (1893, 1896) not only suggested the impact origin of the lunar craters, but made a careful terrestrial study of the one-kilometre-diameter Meteor Crater in Arizona, though he came out against an impact origin in this case (Seddon, 1977). A number of smaller craters or clusters, associated like Meteor Crater with iron meteorite material, were also accepted as meteoritic, from 1928 through the 1930s (McCall, 1977) – Kaalijaarv, Estonia; Wabar, Arabia; Henbury and Boxhole, Australia; Campo del Cielo, Argentina (the latter with a history of iron finds back to 1576, when the natives said they 'fell from the sky'). Dalgaranga, Australia (1938) (McCall, 1965a) was unique in being associated with a stony-iron meteorite find, not an iron. Wolfe Creek, Australia, almost matched Meteor Crater in size, and like it required an impact explosion (McCall, 1965b) (Figures 2a, b and c). Nature proved the point in a crop of more than a hundred craters up to twenty-five metres in

**Figure 2a**. View of Wolfe Creek Crater, Western Australia.

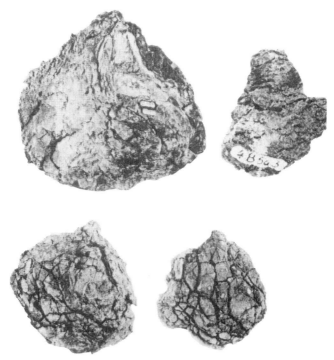

**Figure 2b.** Iron oxide shale ball from within the Wolfe Creek Crater.

**Figure 2c.** Iron (octahedrite) meteorite fragment found three kilometres from it.

diameter in a spectacular event involving jagged masses of nickel-iron in the Sikhote Alin region of Siberia, in 1947. Then, with the space age encouraging search and research, more and more larger-impact explosion craters and circular structures were recognized by geologists across the globe – even two or three under the sea – so that the present count is more than 150 (Grieve, 1998; McCall, 2001a), and they range back in geological time to nearly 2,000 million years. The largest, Chicxulub, Yucatan, is 180 kilometres in diameter, equivalent to an asteroidal or cometary strike.

## GEOLOGICAL/SOLAR SYSTEM TIME AND BACK BEYOND

There are four periods of time that are significant in the history of any meteorite (Hutchison and Graham, 1994): terrestrial age, cosmic-ray exposure age, formation age and formation interval.

*Terrestrial age* is the time spent on Earth since fall. Obviously, the material from an observed fall has an immediately known terrestrial age (zero at the time of fall). Cosmic-ray-induced isotopes are used to obtain the terrestrial age from finds, meteorites that fell from the sky at some time in the past. We know from meteorites analysed immediately after their fall how much of certain isotopes are present in a meteorite when it arrives. A meteorite that is found but not seen to fall will have less of these isotopes because the Earth's atmosphere protected it after arrival on Earth, and unstable products of cosmic radiation such as $^{14}C$ will decay, so that the difference between the normal content of them on arrival and that measured in the meteorite after find can be used to determine the terrestrial age. Because meteorites decay through natural weathering processes, these ages are usually values of no more than tens of thousands of years in temperate regions, but in arid regions, such as the Nullarbor Plain, they are likely to be a bit more, most however being less than 36,000 years, though an exception is about 1 million years in the case of the large Mundrabilla iron (Figures 3a and 3b). In the Antarctic the ages taper off about 300,000 years though a few have ages of 1 million to 3 million years (Figure 4). There is even a unique case of meteorites known to have fallen 480 million years ago in Sweden, though the determination depends on geological evidence, not isotopic evidence – we shall return to this under 'Fossil meteorites'. The

**Figure 3a.** The largest 11-tonne mass of the Mundrabilla iron meteorite resting above the limestone surface, as it came to rest after falling from the sky about 1 million years ago.

**Figure 3b.** The second-largest mass is of 5 tonnes, separated from the limestone surface by a 10-cm pad of iron shale, the product of weathering throughout that time span.

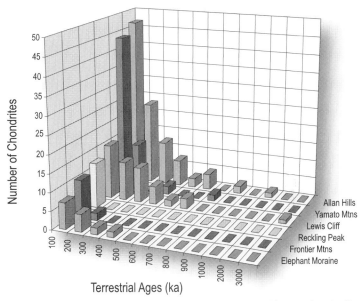

**Figure 4.** Terrestrial ages of about 280 Antarctic meteorites, sorted by stranding site (from Zolensky, 1998).

method is reliable except for the fact that a large meteorite may have its interior shielded from cosmic rays while in space, causing erroneous results. Besides $^{14}$C, the isotopes $^{81}$Kr, $^{36}$Cl and $^{26}$Al are also utilized.

*Cosmic-ray exposure age* is the time spent as a metre-scale meteoroid orbiting the Sun. Cosmic rays react with some atoms in iron or stony meteoroids, and the quantity of isotopic products formed depends upon the chemical nature of the meteoroid and the duration of exposure to cosmic rays in space. The most usual measurements are of the quantity of neon gas resultant on cosmic-ray exposure. The evidence suggests that few stony meteorites survive in space without further collisional pulverization and destruction for more than 40 million years, but iron meteorites, which account for only 4 per cent of falls, are more robust, surviving up to 1,000 million years.

*Formation age* is the age between the last high-temperature episode in the parent body and the present. In the case of basaltic achondrites, this represents the time of crystallization from the liquid in a magma: chondrites, the most common meteorites, which have slightly greater formation ages, did not melt but were hot and recrystallized as solids soon after formation. The method involves the normal radioactive 'clocks' used by geologists, such as uranium–lead, the amount of lead produced by radioactive decay being an indicator of formation age. Values for chondrites are near 4,550 million years; some parent bodies of chondrites were then heated and recrystallized, the primitive chondrite texture being lost: others were heated to the extent that they melted with fractional crystallization during the next 100 million years producing the achondrites with magmatic textures, like igneous rocks. Some chondrites retain very primitive textures and volatiles, including water: the best known of these primitive chondrites are the carbonaceous chondrites. It has been suggested that these come from asteroids in the further part of the asteroid belt, between Mars and Jupiter, where it is colder, and that the common chondrites come from the inner part where it is warmer – this, however, is speculation.

*Formation interval* is the time between the formation of the elements in stars (where almost all the elements except hydrogen and helium were formed) and their incorporation in the parent body. This is done by measurement of decay products of plutonium, an element that, because of its short half-life, is almost unknown surviving on Earth. Plutonium was formed in a star about 150 million years before the formation of the asteroidal parent bodies of meteorites, but other elements were formed at different times.

These are time connotations quite familiar to meteoriticists, but we shall now consider some less well-known aspects of the time relationship.

## METEORITE FINDS FROM FALLS LONG AGO AND THE TERRESTRIAL ENVIRONMENT

A seminal paper by Bland, Bevan and Jull (2000), from which this section is derived, developed a quite novel application of terrestrial ages of meteorites, their weathering (carbonate growth, oxidation). The flux of meteorites to Earth has been constant for the past 50,000 years. Hughes (1981) deduced this from fall statistics and Halliday and Griffin (1982) from entry dynamics of asteroidal debris. The destruction by weathering in hot dry deserts and cold dry deserts such as Antarctica has been much slower than elsewhere, allowing accumulation of meteorites. In Antarctica, where meteorites survive longer than anywhere else because of constant environmental conditions and slow chemical weathering, they accumulate because of burial and ice movement and surface where the ice is arrested by obstacles such as nunataks. In the hot deserts, such as the Nullarbor Plain, they accumulate residually without redistribution, simply because the aridity inhibits rapid weathering.

## WEATHERING PROCESSES AFFECTING METEORITES IN HOT AND COLD DESERTS

Bland et al. considered the application of $^{57}$Fe Mossbauer spectroscopy to constrain some weathering processes in meteorites, such as 'rusting', formation of evaporite minerals and halogen contamination. Any significant Fe+++ in common chondrites, which make up 80 per cent of falls, can be attributed to weathering because olivine, pyroxene, troilite and iron/nickel metal make up 90 per cent of their composition. Mossbauer analysis thus gives an overview of the degree of weathering in the entire meteorite sample. Checks by repeated analyses and on paired samples show that the method is valid. The composition of the original meteorite is an important factor, and H (high-iron) chondrites weather more than L and LL (low-iron) chondrites. Taking this into account, the research depends on terrestrial age and Mossbauer analysis.

X-ray diffraction and scanning electron microscope analysis of evaporitic minerals (carbonates and sulphates) were also taken into account. Antarctic weathering of stony meteorites is a multiple process, involving rusting, hydration of silicates and glass, and hydration and solution of carbonates. Evaporite minerals form rapidly and are composed of mixed terrestrial (C, O, H) and meteoritic elements (Mg, Ca, Na, K, Rb).

Despite the extreme aridity of Antarctic environments, halogen abundances display irregularity. Iodine, in particular, is taken from the atmosphere and there is an overabundance. The source is believed to be $CH_3I$ formed by organisms living on the ocean surface. The atmosphere reacts with rock to incorporate iodine in the meteorites.

## GEOMORPHOLOGICAL APPLICATIONS

Whether meteorites are affected by deflation, buried after fall, or moved away from their site of fall in a particular desert region, are important considerations. Very valuable evidence concerning deflation was obtained in the case of the Mundrabilla iron meteorite masses, distributed in a huge ellipse on the Nullarbor Plain (McCall, 1998), and shown by Aylmer et al. (1988) to be more than a million years old. One piece was found resting on a mound, thirty centimetres above the plain, shielding part of the ground from deflation. The height of the mound gives a lower limit to deflation of the clay cover in this area.

In the reg (stony desert) region of the Sahara, meteorites rest on rock, whereas in the Nullarbor they rest on a thin clay cover to the limestone, yet they do not form craters and only plug in when they fall during rare rainy periods. All the evidence suggests, therefore, that they have not moved after fall, and are found at the site of fall with no significant portion buried. Determinations of meteorite flux by Halliday et al. (1989) and Bland et al. (1998) thus remain applicable to the Nullarbor Plain.

Bland et al. plotted oxidation frequency distribution against terrestrial age, covering ordinary chondrites in the Sahara, Nullarbor, Saudi Arabia, Roosevelt County (New Mexico) and the southern high plains of the USA. This revealed clearly the compositional effect of increased weathering of the H-type as against the L-Y and LL-type ordinary chondrites.

Processes of erosion and volume expansion associated with growth of alteration products gradually destroy meteorites. Larger meteorites do survive longer. Material is gradually removed from the meteorite until it ceases to exist. Bland et al. plotted the frequency of occurrence of oxidation states (Figure 5) for Saharan against other hot desert ordinary chondrites and this showed a peak at 30–35 per cent oxidation compared with 40–45 per cent for other hot desert regions. The plot of degree of oxidation against terrestrial age again separated Saharan from other meteorites found in hot deserts, and showed a very rapid initial phase of alteration and a general increase in weathering over time. This reduction in weathering rate after the initial phase may be due to porosity reduction. They interpreted the shift between the Saharan and other hot-desert regions as indicting relatively low age for

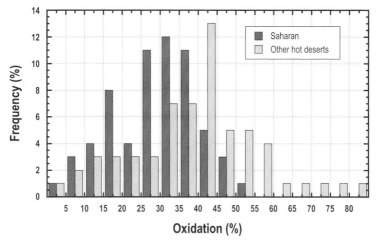

**Figure 5.** Ferric oxidation (expressed as a percentage of total iron and derived from Mossbauer spectra as a frequency distribution) for meteorites from the Sahara (black) and other hot desert areas (grey). The offset of the peaks is explained by a relatively young age for the Sahara accumulation surface (from Bevan et al., 2000). Reprinted from *Quaternary Research*, 53, 2000, with permission from Elsevier.

the surfaces on which Saharan meteorites accumulated, i.e. the difference was related to the timing of climatic changes in different regions. The Nullarbor appears to have had a stable climate conducive to the accumulation of stony meteorites for the past 30,000 years, whereas the Sahara had a stable climate for a shorter period. The results for the Nullarbor were consistent with speleothem-derived evidence for climatic changes, and the conclusion was reached that meteorite-derived evidence helps to define these changes more precisely.

Another plot, of terrestrial age against oxidation state, showed that weathering rates are at least two orders of magnitude slower for Antarctic common chondrites than for meteorites from hot deserts (Figure 6). It is apparent from this highly original research that meteorites can be used to explore the relationships between climate and rock weathering.

These studies of oxidation have also been extended in the Allan Hills, Antarctica, to determine the rate of ice flow (Bland et al., 1998), the results being consistent with a rate of movement of the ice towards the coast of about one metre per year.

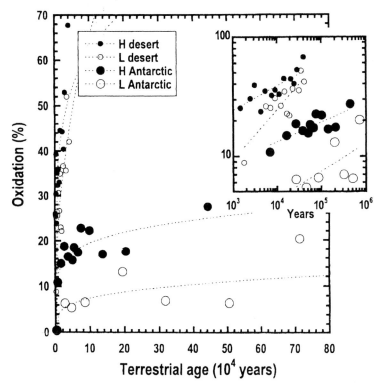

**Figure 6.** Oxidation plotted against terrestrial age for hot desert and cold desert (Antarctic) ordinary chondrites of H and L classes; data binned by terrestrial age (from Bevan et al., 2000). Reprinted from *Quaternary Research*, 53, 2000, with permission from Elsevier.

## FOSSIL METEORITES

The research discussed above concerns meteorites that fell during the Quaternary (Late Pleistocene, Holocene or Recent). For a long time, scientists were surprised that meteorites are not found preserved in really old rocks, like fossils. Not a single one had been discovered in billions of tonnes of coal mined or rock quarried. This led to a discussion in the mid twentieth century as to whether the fall of meteorites to Earth is a relatively recent phenomenon, and there were even arguments advanced that falls of irons and stones did not happen prior to the late Quaternary. Countering this, meteoriticists argued that just as fossil preservation requires very rare taphonomic conditions, so meteorites are an extreme rarity and minuscule in quantity compared with terrestrial rock; discovery would not statistically be expected.

With the advent of isotopic dating methods to determine terrestrial age, we can now take the date of fall on the Earth back to some 3 million years for some Antarctic finds and to 1 million years for the Mundrabilla, Western Australia, iron. However, these are not fossil meteorites, i.e. preserved in rock.

The breakthrough concerning fossil meteorites came in 1952 when a meteorite was found in Middle Ordovician limestone at Brunflo in Central Sweden (63° 04′N, 14° 50′E) (Hutchison and Graham, 1994; Schmitz, 2003). The 10-centimetre-diameter black clast in reddish limestone was not identified as a fossil meteorite until 1979. During the 1980s another such find was made at the Thorsberg quarry, Kinnekulle, in southern Sweden (58° 35′N, 13° 26′E). This, named Osterplans, was in similar strata but about 5 million years older. A systematic search in this quarry between 1992 and 2000 yielded 40 fossil meteorites, and since then 15 more have been found (Schmitz, 2003). The 40 meteorites are at six different stratigraphic levels and represent repeated falls over about 2 million years. They are all L-chondrites and although much altered, being replaced by clay minerals, calcite and barite (both on the sea floor and diagenetically after burial; the masses also have a bleached rock halo), some chondrules retain their outlines (Figure 7). The period covered by the falls is well constrained by two conodont biozones and known average sedimentation rates for this type of orthoceratite limestone. It is estimated from these calculations that at this geologically ancient time the meteorite flux was two orders of magnitude greater than the present flux. This was apparently a repeated rain of meteorites and it is considered to represent the break-up of a meteorite parent body (asteroid) in space 480 million to 500 million years ago. This evidence would seem to show that stony meteorites were raining on the Earth again and again during Early Palaeozoic time.

However, there is a further exciting implication. The meteorites appear about twelve times in different sedimentary horizons, according to latest reports. Modern meteorite falls do not repeat themselves at the same locality for the simple reason that the Earth is rotating on its axis, and this applies equally to the Ordovician as to the present day.

Sedimentary reworking can be eliminated as a possibility in the case of a limestone of this sort; repeated narrowly targeted falls again and again at the same site over millions of years stretch the limits of credibility, so the conclusion must be reached that a large part of the globe was sprinkled again and again during this period. There may be a relationship to the recurrent Leonid type of meteor shower, but these only rain minute meteoroids that burn up in the atmosphere, not a rain of stones. We are dealing with repeated meteorite rainings of cobble- and boulder-sized objects that must have covered a very large area of the globe: though they are unlikely to have been panglobal, for orbital reasons, they may well have covered the circumference of the turning globe. The Brunflo and Osterplans sites are more than 500 kilometres apart (Figure 8).

Geologists should search carefully at equivalent Llanvirn–Arenig deposits elsewhere, and especially limestones, because they apparently exert a peculiar preservational influence on these, in other circumstances ephemeral because of rapid decay by weathering, meteoritic bodies.

**Figure 7.** A chondrite fragment in the red orthoceratite limestone at the Osterplans Quarry (from Schmitz, 2003). The scale is obvious from the hammer.

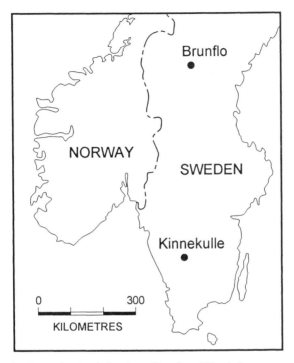

**Figure 8.** Sketch map showing the locations of Brunflo and Kinnekulle.

## LUNAR-SOURCED METEORITES AND TIME

In 1964 and 1965, I attended two NASA-sponsored meetings in the USA with the aim of predicting what the lunar surface would be made of when the first Apollo mission landed on it. No one there in their wildest dreams suggested that pieces of it were lying about the surface of our planet – but they were! Would we have recognized them as such, if we had recovered and examined any before Apollo 11 landed in 1969? I doubt it. In fact, it was not until 1981 that meteorite ALH 81005 from the Allan Hills in Antarctica was recognized as a lunar upland anorthosite breccia. Several more lunar achondrites have been recovered in Antarctica since then; another has been recognized in Australia (a specimen already in a collection, recorded as found amidst the spread of masses from the 1960 Milly Milly achondrite fall in

Western Australia, was identified as lunar and is now renamed Calcalong Creek (Hill et al., 1991)); and there are two more from Libya, where separate lunar meteorites and another believed to come from Mars were found within an area fifty kilometres in diameter (McCall, 1999). The largest lunar meteorite so far recovered on Earth is Dar-al Ghani 400 from Libya, which weighs 1.425 kg. The mind boggles at the implications of this proximity to the amount of lunar meteorite material that has sprinkled the Earth in geologically recent time.

The same isotope-based methods described above for asteroidally sourced meteorites can be applied to lunar meteorites, and none of the lunar meteorites found has a geologically long terrestrial age. Of seventeen such finds, terrestrial ages range from about 20,000 to 200,000 years, whereas cosmic-ray exposure ranges (which may represent a combination of transit time from the Moon and un-protected exposure in the lunar regolith) have a wider range, but most are between 100,000 and 15 million years). These lunar meteorites are all from familiar surface rocks, mare basalts, diabase, gabbro; impact melts; highland anorthosites or mixtures of these; and most are breccias similar to the surface breccias collected by Apollo astronauts and the Russian Luna unmanned mission from the lunar surface – though the anorthosites come from the far side of the Moon not explored other than instrumentally, according to matching with remote instrumental data collected by orbiting Apollo spacecraft. The ages obtained from the isotopic evidence supplied by these finds indicate that they are all products of impact ejection to space from the lunar surface during late periods of lunar history, when such impacts were relatively small. It must be accepted, therefore, that even such small-scale impacts, after the great early bombardments (widely invoked to explain why the large-scale craters ceased) were apparently sufficiently powerful to send solid rock masses to impact on the Earth.

I find the story of the lunar-sourced meteorites above amazing enough, yet it is even more surprising to look back figuratively in time to what planetologists widely believe went before, and from when we do not have any material recovery. The concept of the 'Great Bombardment of the Moon' around 4,000 million to 3,900 million years ago is firmly entrenched. Many scientists like myself in the 1960s seriously entertained the idea of an internal, eruptive origin for the lunar cratered surface (e.g. McCall, 1965c), but the opposing impact hypothesis won the day with NASA. However, the scale of the largest craters and structures on the Moon attributed to the early 'Great Bombardment' must surely – if they were so formed – require that immense amounts of material from deep below the lunar surface were ejected to space during these events.

There is no trace at all in the Earth's rocks of this lunar material or of the direct bombardment which the nearby Earth, like the Moon, must have suffered, but this is perhaps understandable because rocks of this great age are now exposed only in quite small areas of the globe, though explanations that all trace has simply been weathered off through the long time that has elapsed are less easy for geologists to accept.

The material ejected at this time from the Moon should have collided with some asteroidally sourced meteorites and formed breccias with them. There is no trace of these. The cratered surface of Mercury (Cross and Moore, 1977; McCall, 2003), if formed likewise, must also have been the site of ejection of immense volumes of rock material into space, though one can explain its apparent absence by the near-Sun position of this planet. Mars and the heavily cratered satellites, some of planet dimensions, of the outer planets may also have spalled off large volumes of rock material.

One can escape from this dilemma conveniently for all but the Moon by saying that either the orbits of asteroids were such that there was no collision or that we have not sampled those asteroids – it is true that we mainly sample meteorites from the near-Earth asteroids and have not sampled the bulk of the asteroids (Zolensky and Ivanov, 2001 list the D, F, P, T and R class asteroids as not spectrographically matched with known meteorite types). One can also escape by saying that all this material was entirely vaporized in the giant impact explosions. However, the first escape route appears unlikely in view of the numerous near-Earth asteroids with Earth/Moon-passing orbits, and the second unbelievable if lumps of the Moon and Mars can have been ejected later in Phanerozoic times by much weaker and fewer impacts to reach the Earth, as most meteoriticists believe: these were clearly projected by impact explosions such as can clearly project rocks into space as well as vaporize target materials.

My argument may be encapsulated in the question: Isn't there too much asteroidal meteoritic material in space and not enough impact ejecta sourced in the Moon, Mercury and Mars – not to mention the cratered satellites of the outer planets?

The corollary to this question is that endogenous processes that do not necessarily involve ejection may yet prove to have had a much greater role in producing the numerous cratered surfaces that pop up wherever space missions explore. The Mariner 10 coverage of Mercury appears to me to reveal many structures quite impossible to explain by invoking impacts, even allowing for a hypothetical and unsubstantiated secondary post-impact deformation process (McCall, 2003).

## SNC (MARTIAN?) METEORITES AND TIME

As with lunar meteorites, the same isotopic dating methods as are applied to the asteroidally-sourced, overwhelming majority of meteorites that fall to Earth can be applied to the SNC (shergottite/ nakhlite/chassignite) meteorites, a group numbering twenty-seven (not including thirteen pairings), which appear to come from Mars. Bridges (2003) has recently presented an up-to-date summary of these, noting that the oxygen isotope compositions and oxidation states are similar to terrestrial igneous rocks (contrasting with the iron/nickel metal occurrence in asteroidally sourced meteorites), and the shergottites have gas inclusions in their shock-melted glass identical with the Martian atmosphere, as sampled by the Viking lander. Some SNCs have evaporite-like hydrated alteration minerals. They have clearly been ejected from a planet on impact – the plagioclase in the shergottites is converted to maskelynite by the shock of the impact. Ejection from Venus or the Earth is ruled out by virtue of their thick atmospheres, Mercury is an arid, long-inert waste like the Moon, and the formation ages for these meteorites of between 4,500 million and 160 million years indicate a source in an active planet. Occam's razor suggests that there is no other possible source but Mars.

The 27 separate meteorites number 19 shergottites (9 basaltic, 5 olivine-rich and 5 lherzolitic – the latter peridotites with dominant olivine + orthopyroxene + clinopyroxene) – 6 nakhlites (clinopyroxenites), 1 chassignite (dunite, entirely composed of olivine) and 1 unnamed orthopyroxenite, the renowned ALH 84001 from Antarctica, supposed (by some) to contain traces of microscopic life forms. The basaltic shergottites have formation ages ranging from 474 million to

165 million years (that is, equivalent to Earth's mid Ordovician to mid Jurassic). The lherzolitic shergottites have formation ages of 180 million years (again equivalent to our mid Jurassic). The nakhlites and chassignites have formation ages of 1,300 million years, equivalent to our Precambrian (Mesoproterozoic). The single orthopyroxenite, ALH 84001, has a formation age of 4,500 million years, not widely different from that of the ordinary chondrites, older than any terrestrial rocks.

These ages can be speculatively correlated with sequential schemes derived from analyses of Martian images for the various terrains on the planet, but in truth these SNC meteorites constitute a remarkably restricted array of rock types – only six – and a restricted range of formation ages is also represented, when one considers the visible complexity of the planet's surface. Why are there no sediments represented if Mars has water- and wind-deposited sediments? Even the igneous rock range is remarkably restricted for an entire complex planetary surface.

## ASTEROIDAL METEORITES AND TIME

The formation ages of asteroidally sourced meteorites, the overwhelming majority, are all close to 4,550 million years, with chondrite ages slightly greater than the basaltic achondrites that were heated, melted and recrystallized to familiar igneous rock textures almost immediately after accretion. There is no problem with such ages, which means that the source asteroids were all formed by accretion early in the history of the Solar System and some suffered heating with some melting just afterwards. I do, however, find a problem with the cosmic-ray exposure (CRE) ages. Hutchison and Graham (1994) provided a lucid account of this method and concluded that evidence from neon isotopes shows that stony meteorites survived as individual meteoroid bodies in space only for periods of a few million years to a few tens of million years, being destroyed by collisional pulverization after a general maximum of 40 million years. They illustrated the Tysnes Islands, Norway, chondrite with a CRE of 5 million years, as an example. Iron meteorites are apparently more robust and can survive in space for up to 1,000 million years and they illustrated this with Picacho, New Mexico, with a CRE of 700 million years.

I find the evidence for stony meteorites surprising. The asteroidal parent bodies were stationed there in space 4,550 million years ago and have been circling round in their orbits, albeit with many collisions, since then. A simple calculation shows that for the stony meteorites there must have been at least the equivalent 114 cycles of collision, production of fragmental meteorite masses and their total destruction. Of course, it did not happen in such regular cycles – the process of collision and spawning of meteoroids and their destruction was a continuing process, but this is a useful quantitative expression of the repetition of collision, orbiting and destruction that is inferred. One would have thought that these repeated collisions and destructions would, in the vast period of time applicable, have exhausted the supply of material. There seems to be a stark contrast between the survivability of asteroidal material, and the complete unsurvivability of the ejectamenta during the hypothetical 'Great Bombardment of the Moon', for which the expectation would also be repeated collisions with near-Earth asteroids and meteoroids sourced from them. The lunar-sourced ejectamenta would surely be no less survivable than the asteroidally sourced meteoroid material.

## CONCLUSION

My essay on time in relation to meteorites has had five strands: new research relating falls to asteroidal parent bodies; the history of scientific acceptance; some new and surprising applications of meteoritics to terrestrial weathering, climates and glacial processes; drawing attention to an at present unique-in-the-record piece of evidence of a 'non-uniformation' series of meteorite-rain events in the distant geological past 480 million years ago in the Ordovician; and showing some awkward questions that still exist despite our remarkable advances in knowledge during the last half century – the 'space age'.

I am by nature of a sceptical turn of mind, and have long followed Lichtenberg's dictum, 'Question everything once.' The achievements in the half-century of the space age of scientists, engineers and technological inventors have been astonishing. Yet there appears to be a problem in space science research concerning the development of entrenched assumptions – the passage from hypothetical models and speculations to entrenched assumptions that with the passage of time come to be treated as fact. I have previously questioned the 'Giant Impact' hypothesis for the origin of the Earth–Moon system (McCall, 2000) and the statistics of 'doom' predictions relating to bodies from space hitting the Earth (McCall, 2001b). My lingering doubts concerning the 'Great Bombardment of the Moon' have resurfaced on account of preparation of a segment on the planet Mercury for an *Encyclopedia of Geology* (McCall, 2003). I find that I am by no means yet convinced that the craters on Mercury can all be dismissed as caused by impact (even with the larger ones subsequently modified by some obscure surface process). Claudio Viti-Finzi (2002) recently cited Runcorn's remark in 1974 that the volcanic alternative for lunar cratering was preferred by geologists 'presumably not wishing to allow catastrophism, expelled by Lyell from their science, in by the backdoor of lunar science'. In fact, advocates of lunar volcanic cratering in the 1960s worked by careful analysis of the lunar structures, of which many of the larger craters do closely resemble volcanic calderas (McCall, 1965c), and had no such thought in mind. These researchers were not proved entirely wrong in their interpretations by subsequent events – I received a very polite letter from H. C. Urey prior to Apollo 11: 'Surely, Dr McCall, you don't believe that the lunar surface is covered with lava flows?' – I regret now that, after Apollo 11 did reveal just that, the letter disappeared, a victim of 'untidy office erosion'.

Let me make it clear, no one argues about the planetesimal accretion theory of planetary origin. There is, however, no real analytical justification other than hypothesis for the assumption that all the lunar craters are impact-generated, but display complex patterns not compatible with impact-cratering theory, modelling or the patterns shown by the 150 or so impact structures now known on Earth (Grieve, 1998; McCall, 2001a), because they have been degraded by some mysterious secondary process. At present such questioning is likely to be dismissed as 'cranky', but, some time after I am gone from the scene after a long innings, these doubts will, I predict, resurface and have to be confronted. There are some significant anomalies in the record of meteoritics that have yet to be explained.

## REFERENCES

Aylmer, D., Bonnano, V., Herzog, G.F., Weber, H., Klein, J. and Middleton, R. (1988), 'Al and Be production in iron meteorites', *Earth and Planetary Science Letters* 88, 107.

Bevan, A.W.R., Bland, P.A. and Jull, A.J.R. (1998), 'Meteorite flux in the Nullarbor region, Australia', in Grady, M.M., Hutchison, R., McCall, G.J.H. and Rothery, D.A. (eds), *Meteorites: flux with time and impact effects*, Geological Society, London, Special Publication No. 140, 59–73.

Bland, P.A., Bevan, A.W.R. and Jull, A.J.T. (2000), 'Ancient meteorite finds and the Earth's surface environment', *Quaternary Research* 53, 131–142.

Bland, P.A., Conway, A., Smith, T.B., Berry, F.J., Swabey, C.T. and Pillinger, C.T. (1998), 'Calculating flux from meteorite decay rates: a discussion of the problems encountered deciphering $10^5$-$10^6$ year integrated meteorite flux at Allan Hills and a new approach to pairing', in Grady, M.M., Hutchison, R., McCall, G.J.H. and Rothery, D.A. (eds), *Meteorites: flux with time and impact effects*, Geological Society. London, Special Publication No. 140, 4–58.

Bridges, J. (2003), 'Starry Messengers', *Geoscientist* 12, 4–9.

Cross, A. and Moore, P. (1977), *The Atlas of Mercury*, Mitchell Beazley, London.

Fothergill, B. (1974), *The Mitred Earl: An Eighteenth-Century Eccentric*, Faber & Faber, London.

Gilbert, G.K. (1893), 'The Moon's face', *Bull. Phil. Soc. Wash.* 12, 241.

Gilbert, G.K. (1896), 'The origin of hypotheses, illustrated by the discussion of a topographic problem', *Science* (New York), new ser. 3, 1–13.

Grady, M.M. (2000), *Catalogue of Meteorites, with special reference to those in the collection of the Natural History Museum, London*, Cambridge University Press, Cambridge.

Grieve, R.A.F. (1998), 'Extraterrestrial impacts on Earth: the evidence and the consequences', in Grady, M.M., Hutchison, R, McCall, G.J.H. and Rothery, D.A. (eds), *Meteorites: flux with time and impact effects*, Geological Society, London, Special Publication No. 140, 105–31.

Halliday, I. and Griffin, A.A. (1982), 'A study of the relative rates of meteorite falls on the Earth's surface', *Science* 23, 31–46.

Halliday, I., Blackwell, A.T. and Griffin, A.A. (1989), 'The flux of meteorites on the Earth's surface', *Meteoritics* 24, 173–8.

Hill, D.H., Boynton, W.V. and Haag, R.A. (1991), 'A lunar meteorite found outside the Antarctic', *Nature* 352, 614–16.

Hughes, D.W. (1981), 'Meteorite falls and finds: some statistics', *Meteoritics* 16, 269–81.

Hutchison, R. and Graham, A. (1994), *Meteorites*, Natural History Museum, London.

McCall, G.J.H. (1965a), 'New material from, and a reconsideration of, the Dalgaranga meteorite and crater', *Mineralogical Magazine* 35, 476–87.

McCall, G.J.H. (1965b), 'The caldera analogy in selenology', in Whipple, H.E. (ed.), 'The caldera analogy in selenology', *Annals of the New York Academy of Science* 123 (Art. 2), 843–75.

McCall, G.J.H. (1965c), 'Possible meteorite craters – Wolf Creek, Australia and analogs', in Whipple, H.E. (ed.), *Annals of the New York Academy of Science* 123 (Art.2), 970–98.

McCall, G.J.H. (1967), 'The progress of meteoritics in Western Australia and its implications', in Moore, P. (ed.), *1968 Yearbook of Astronomy*, 146–55, Macmillan, London.

McCall, G.J.H. (1973), *Meteorites and their Origins*, David and Charles, Newton Abbot.

McCall, G.J.H. (1977), *Meteorite Craters*, Dowden, Hutchinson & Ross, Stroudsburg, Pennsylvania.

McCall, G.J.H. (1998), 'The Mundrabilla iron meteorite – an update', in Moore, P. (ed.), *1999 Yearbook of Astronomy*, 156–68, Macmillan, London.

McCall, G.J.H. (1999), 'Meteoritics at the millennium', in Moore, P. (ed.), *2000 Yearbook of Astronomy*, 153–77, Macmillan, London.

McCall, G.J.H. (2000), 'The Moon's origin – constraints on the Giant Impact Theory', in Moore, P. (ed.), *2001 Yearbook of Astronomy*, 212–17, Macmillan, London.

McCall, G.J.H. (2001a), 'Tektites in the geological record: showers of glass from the sky', *Earth in View: No 1*, Geological Society Publishing House, Bath.

McCall, G.J.H. (2001b), 'Keep watching the skies – but not in fear', *Geoscientist* 11 (3), 12, 17.

McCall, G.J.H. (2003), 'Mercury', in Selley, R., Cocks, L.R.M., McCall, G.J.H. and Plimer, I.R. (eds), *Encyclopedia of Geology*, Elsevier, Kidlington, Oxfordshire.

Schmitz, B. (2003), 'Shot stars', *Geoscientist* 13 (5), 4–7.

Seddon, G. (1977), 'Meteor Crater – a debate', in McCall, G.J.H. (ed.), *Meteorite Craters*, Dowden, Hutchinson & Ross, Stroudsburg, Pennsylvania, 157–69.

Viti-Finzi, C. (2002), *Monitoring the Earth*, Terra, Harpenden.

Zanda, B. and Rotaru, M. (2001), *Meteorites: Their Impact on Science and History*, Cambridge, University Press, Cambridge.

Zolensky, M.E. (1998), 'The flux of meteorites in Antarctica', in Grady, M.M., Hutchison, R., McCall, G.J.H. and Rothery, D.A. (eds), *Meteorites: flux with time and impact effects*, Geological Society, London, Special Publication No. 140, 93–104.

Zolensky, M.E. and Ivanov, A.B. (2001), 'Kaidun: a smorgasbord of new asteroid sample', *Meteoritics and Planetary Science* 36 (9), Supplement, A233.

# Porrima: A Close Approach

BOB ARGYLE

## INTRODUCTION

Visual binaries, to the small-telescope user, are pairs of stars that are physically connected and rotate around the common centre of gravity in periods ranging from tens to many thousands of years. This motion manifests itself in a change of angle between the two stars (position angle) and a variation in the angular separation of the two stars with time. Position angle is reckoned in the following manner. When the fainter of the two stars is due north of the primary the position angle is 0°. When it is east the position angle is 90° and so on through 180° (south) and 270° (west). Separation is measured in arcseconds.

As a general rule, the closer the pair the shorter the period, and those with telescopes of around twenty centimetres or so will find only a handful of bright binaries in which the change in orientation (or position angle) and separation is noticeable without recourse to a piece of measuring equipment such as a filar micrometer. Even where the period is relatively short (fifty or sixty years), the change in position angle is not usually significant on timescales of less than a year or so. Careful estimates can show angular movements of 10° without difficulty, and separation changes can be monitored in terms of the angular diameter of the stellar disks.

In double-star measurement, a third parameter is needed along with the position angle and separation. This is the date (or epoch) of observation. It is always given in decimals. One day is approximately 0.0027 years, so 1 February 1836 for instance is 1836.09. Most visual observers use two or three decimal places – measures of short-period binaries with modern techniques usually warrant four decimal places.

If a bright pair were to be followed for the whole of its orbital period, and the position angle and separation were plotted (regarding the brighter star as fixed), the result is known as the apparent ellipse. The shape of this ellipse can vary enormously because it represents a projection on to the plane of the sky of the true ellipse. The speed at which the companion appears to rotate about the primary also depends on the eccentricity (the degree of ellipticity) of the true ellipse. The greater the eccentricity, the faster the companion will appear to move as it passes through periastron (the closest approach between the stars). With increasing aperture the possibility of seeing a visual binary change its position is increased. Christopher Taylor (see Argyle, 2004) suggests that at the correct orbital phase it is perfectly possible to see a rapid change in apparent position angle and separation in the 5.7- year visual binary Delta Equulei over a period of eight weeks using 12.5-inch reflector, but such a feat requires excellent optics, a well-collimated telescope and good seeing.

Very rarely, the components of bright binary stars of long period will make close approaches to each other where the apparent motion is obvious over a matter of days rather than years. Just such a case is Gamma Virginis or Porrima, a name that refers to the Roman goddess of prophecy. Another name for the star, often used, is Arich.

## HISTORY

The exact circumstances of the discovery of Porrima seem to be in some doubt. Lewis (1906), in his section on the star, states that it was first noted as double by Father Richaud in Pondicherry in 1689. However, in the introduction to the same book and, further, in an article in *Observatory* (1908) he acknowledges Bradley and Pound as the discoverers. In 1718 James Pound, observing at Wanstead, noted on 11 March that 'the direction of the double star (viz. gamma) of Virgo was parallel to a line through alpha and delta' which Bradley confirmed the following night. It is interesting that Pound says *the* double star of Virgo as if he was not surprised to happen upon it. Had someone already noted the pair as double? There are no references to an earlier observation. Knowing that Flamsteed had discovered the wide pair in the Zeta Cancri triple system at about 6 arcseconds separation, a check was made in his Catalogue (1725). From 14 February 1609 to 27 December 1719 he observed the star seventy-three times but with no comment on duplicity. Using the most reliable modern orbit the pair would have been separated by 3.76 and 5.48 arcseconds at the two epochs in question. In 1720, Cassini observed both the occultation and re-emergence from the lunar limb of the stars but the first reliable micrometer results were not made until well into the nineteenth century.

The true orbit of Porrima is a remarkably elongated ellipse. The eccentricity is almost 0.9, making it one of the highest values among the bright visual binaries, and the apparent separation varies by a factor of 17, reaching 6 arcseconds at widest. In the true orbit the distance between the two stars varies from 4.7 AU (astronomical units) at periastron to more than 82 AU at apastron. The widest separation in the apparent orbit occurred in 1920 and will happen again in 2089. The inclination of the true orbit is about 20° to the line of sight, so there is not much foreshortening involved and what we see is close to the actual situation. Another consequence of the orbital eccentricity is that it takes about 165 years to cover the 180° of position angle including apastron, while it only take five years to cover the opposite half of the apparent orbit. At closest approach the position angle will be changing at the rate of 1° every five days or so.

## THE FIRST ORBITS

In 1828 the Frenchman Savary had, for the first time, produced the orbital elements for a binary star, Xi UMa, from measurements of the position angle and separation. During the early 1830s it was clear that Porrima was going to make a close approach sometime in the decade, and as a prelude to this John Herschel studied the possible orbital solutions for the star using the available observations. Herschel's first attempt to solve the problem of Porrima in 1831 was the second such time this was done. Dissatisfied with this, he later made another attempt to derive the orbit in 1833.

In both cases the periods were well over 500 years. From the second orbit, Herschel had expected that the closest approach would be 0.51 arcseconds and this would occur in 1834.39. However, the star did not cooperate and he recalculated the orbit, this time with a longer period and with a periastron passage occurring at 1834.63. Again this time came and went and Porrima was still a relatively easy object, F.G.W. Struve having measured the separation at 0.91 arcseconds a few months earlier. However, in late spring 1835, Struve found that the distance had reduced to 0.51 arcseconds so it was clear that periastron was imminent. Herschel blamed his problems on the observation by Pound and Bradley in 1718. He thought it was totally incompatible with any reasonable ellipse. Unfortunately, Herschel had made an error amounting to 10° in calculating the position angle that Pound and Bradley had implied from their observations.

## THE OBSERVATIONS

Table 1 gives a selection of observations made between 1834 and 1838 and covering the last close approach. The main observers were Dawes, Smyth and F.G.W. Struve. Herschel himself would only make estimates, although his negative observation of February 1836 is one of the most significant contributions.

The data in the following table is a modified version of the measures in the WDS Observations Catalogue between 1834.0 and 1838.0 – some have been combined into a single mean and position angles have been altered by 180° to bring them into conformity. In the following table all observations were made with apertures of 10 inches or less and in Dawes's case only 4 inches. The distance estimates by F.G.W. Struve between 1836.40 and 1836.42 must be underestimates if we accept that Dawes and Smyth did see an elongation a month or so previously, particularly as they were using smaller apertures than Struve. The Dorpat observations were made on three nights in May 1836 by F.G.W. Struve, his son Otto who was 17 years old at the time, and Georg Sabler (1810–64), one of Struve's assistants, who does not appear to have had much experience in double-star observation. Struve alone made estimates of separation based on the ellipticity of the elongated image and in each case a power of 1,000 was used on the 9.6-inch-aperture Fraunhofer refractor (Figure 1). The other two observers made measures of position angle only.

**Table 1. Some Selected Micrometric Measures of Porrima**

| Epoch | PA (°) | Sep. (″) | Nights | Observers |
|---|---|---|---|---|
| 1834.30 | 229.0 | 0.91 | 15,10 | Dawes, Smyth, Struve |
| 1835.38 | 195.4 | 0.51 | 11 | Struve, Senff, Smyth |
| 1836.06 | | round | | Smyth |
| 1836.13 | | single | | Herschel |
| 1836.15 | | round | | Smyth |
| 1836.28 | 169.6 | | 2 | Dawes |
| 1836.34 | 169.8 | | 2 | Smyth |
| 1836.40 | 156.1 | | 1 | Otto Struve |
| 1836.41 | 151.6 | 0.26 | 3 | F. G. W. Struve |
| 1836.41 | 153.8 | | 1 | Sabler |
| 1836.42 | 161.3 | | 1 | Otto Struve |
| 1836.59 | 116.3 | | 3 | Encke, Mädler |
| 1837.20 | 100.3 | | 1 | Encke |
| 1837.21 | 85.4 | 0.6 | 1 | Smyth |
| 1837.44 | 75.6 | 0.65 | 18 | Encke, Struve |
| | | | | |
| 2001.26 | 253.2 | 1.24 | 3 | Alzner |
| 2001.31 | 252.3 | 1.27 | 8 | Argyle |
| 2002.31 | 242.1 | 1.06 | 8 | Argyle |
| 2002.36 | 243.7 | 1.04 | 6 | Alzner |
| 2003.34 | 232.2 | 0.83 | 5 | Alzner |
| 2003.35 | 231.5 | 0.87 | 7 | Argyle |
| 2004.23 | 212.9 | 0.63 | 4 | Alzner |
| 2004.29 | 209.1 | 0.67 | 4 | Argyle |

## THE OBSERVERS

Friedrich Georg Wilhelm Struve was born in Altona in 1793, one of a family of fourteen children of whom seven died in infancy. He studied at the University of Dorpat in Estonia and was offered a job teaching history at the Dorpat Gymnasium. However, he came under the influence of the professor of physics, George Parrot, turned to astronomy, and was effectively in charge of the observatory at the university from 1813. By 1814 he was using the transit instrument to measure double stars and was able to confirm the elder Herschel's observations

**Figure 1.** The Fraunhofer telescope at Dorpat (now Tartu) Observatory. (By courtesy of Taavi Tuvikene, Tartu Observatory.)

of the motion of Castor by showing that the two components were moving around their common centre of mass. When Struve heard that Fraunhofer was building a refractor with an aperture of 9 Paris inches (9.6 inches) he immediately felt that this telescope should come to Dorpat. This it did in November 1824, and by 1827 Struve had re-surveyed all the known double stars down to −15 degrees and found more than 2,500 of his own. This involved examining stars at the rate of several hundred per hour. The *Catalogus Novus*, which summarized this work, won Struve the Gold Medal of the Royal Astronomical Society. Between 1827 and 1839 Struve continued to make many measurements of the pairs in the *Catalogus* and produced the *Mensurae Micrometricae* in 1837.

Admiral William Smyth is best known for his book *Cycle of Celestial Objects*. Having retired from the Royal Navy he set up a well-appointed private observatory at Bedford in 1830 with a 5.9-inch Tulley refractor and carried out double-star observations with it. In 1844, he published *Cycle of Celestial Objects*, which was followed by a second volume called *The Bedford Catalogue* and a supplementary book to this commonly known as *Speculum Hartwellianum* in which he recounts the history of Porrima.

The Rev W.R. Dawes was a skilled double-star observer with exceptional eyesight (he was also known as the 'eagle-eyed'). He used a variety of telescopes up to and including an 8.25-inch Alvan Clark refractor. His last telescope, with an 8-inch OG by Cooke, is now at the observatories at the University of Cambridge and is used regularly by the writer, who can attest to its excellent imaging properties. Dawes drew up the famous Dawes limit – an empirical formula for the resolution of a telescope as a function of aperture that is an excellent approximation for small telescopes.

## CLOSE APPROACH

The proximity of the two stars in early 1836 seems to have caught everyone by surprise. John Herschel, writing in the *Monthly Notices* of the RAS, reports the following observation made with his 20-foot (18-inch) reflector from Feldhausen at the Cape of Good Hope.

Gamma Virginis . . . at this time is to all appearance a single star. I have tormented it, under favourable circumstances, with the highest powers I can apply to my telescopes, consistently with seeing a well-defined disc, till my patience has been exhausted; and that lately, on several occasions, whenever the definition of the stars generally, in that quarter of the heavens, would allow of observing with any chance of success, but have not been able to procure any decisive symptom of its consisting of two individuals.

On 17 February, he records that Gamma Centauri is seen as double 'without difficulty' (then separated by about $0''.72$) and Saturn was seen with 'uncommon distinctness', but when he pointed the telescope at Porrima, 'it bore a magnifying power of 480 with sufficient distinctness but without indicating the slightest elongation or giving any symptom of its being otherwise than a single star. Had the centres of the two stars been only half a second asunder, I think I could not have failed to see a division between them.'

Another observer who awaited the close approach was Admiral Smyth, and in January 1836 he turned his 5.9-inch refractor on it and found it single. He says, 'This apparent state of singleness may have existed during the latter part of 1835, for when I caught it . . . it was very near a change.' In the *Bedford Catalogue* he shows the appearance of the star. The news that the star appeared single did not seem to engender much interest. Smyth wrote, 'I instantly announced this singular event to my astronomical friends, but the notice was received with less energy than such a case demanded.'

Even more interesting is that, with his 5.9-inch refractor and powers up to ×1,200, he sees an elongation at about 170° in 1836.39, i.e. almost contemporaneous with the measures at Dorpat. What is more, a measure made about three weeks earlier by Dawes bears this observation out. The conclusion to be drawn from this is that closest approach had already occurred by the time Smyth and Dawes made their observations in spring 1836 and that the Dorpat observations must be treated with a great deal of caution. The critical observation is that of Herschel in February 1836. The star appeared single to him on a night when a separation of 0.7 arcseconds was easily distinguished. He was using a 20-inch telescope from a site where Gamma Virginis would be no more than about 34° from the zenith at culmination, not the 58° or so it is from Dorpat (now Tartu in Estonia). This surely suggests that at that time Gamma Virginis was not more than 0.3 arc-seconds apart – perhaps less. This would certainly explain why Smyth saw the star as round at the same time and also why Smyth and Dawes were able to see an elongation two months later, and hence by the time the Dorpat observers had measured it the pair was already widening again. It remains to be seen if this is confirmed by the events of 2005.

## CURRENT ORBITS

Now the events of 1835–6 will be repeated again in spring 2005, although it is not yet certain when the closest approach will be and, as we have seen, the exact distance at this time still seems to be uncertain. In the last few years, as the companion gathers pace, a number of orbits have been computed. There are currently three in the literature with periods ranging from 168.68 to 168.93 years. The most recent, that computed by Soderhjelm (168.9 years), agrees best with recent observations. Figure 2 shows the recent observations made by Andreas Alzner (using a 32.5-cm Cassegrain and Meca-Precis double-image micrometer) and the writer with the 8-inch OG at Cambridge and RETEL micrometer for each year from 2000 to 2004, superimposed on the theoretical curve of Soderhjelm's orbit.

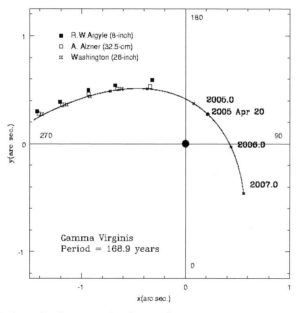

**Figure 2.** The predicted apparent orbit of Porrima from 2000 to 2007. Periastron passage is expected on 20 April 2005.

## VARIABILITY?

Whereas Porrima appears to be a pair of equally bright stars of spectral class F0V, some questions have been raised about the variability of one or other of the components. The earliest suspicion of variability goes back to 1818 when F.G.W. Struve thought the preceding star slightly the fainter. The reference for this is Agnes Clerke, as quoted in Lewis's fine volume (1906). The quote continues: 'It continued so for some years, when from 1825 to 1831 they became again equal, and in 1834 the preceding star was brighter. Otto Struve from 1840 to 1853 noted oscillations of half a magnitude in a few days, but finally decided that star was too low for him to determine the period. From 1840 to 1847 Dawes found each alternately about a quarter magnitude brighter than the other. Dembowski also noted similar swayings of lustre.'

John Herschel made estimates of the brightness of Porrima using the brighter stars of Corvus as comparisons during his observation of February 1836 when the star appeared single. The orbit is clearly far removed from edge-on and so the idea of a drop in brightness due to occultation would only apply if the two stars were immensely large and very close to each other, i.e. within a few stellar radii. In fact, at a distance of thirty-nine light years an F0V star, which has a diameter about 1.3 times that of the Sun, subtends an apparent diameter of just one milli-arcsecond. At that time no stellar distances were known (it would be another two years before Bessel announced the parallax of 61 Cygni) and nothing was known about stellar diameters apart from that of the Sun.

Krisciunas and Handler suggested that it might be a member of the Gamma Doradûs class of pulsating stars. Measurements of the system brightness using photoelectric techniques showed no evidence of variability. Abt and Golson measured a maximum amplitude of just 0.03 magnitudes in V, while the Hipparcos satellite records a maximum range in the magnitude of the combined light of just 0.01 in V.

## THE NEXT CLOSE APPROACH

Observations of Porrima over the next three or four years will be most useful and, whereas accurate measures of position angle and separation will be vital, for those who wish to enjoy the spectacle it should be possible to follow the change of position angle with either a single wire fitted in the eyepiece or a graticule eyepiece of some variety. At minimum separation it seems likely that at least a 12-inch aperture will be necessary to resolve the pair and possibly more. For more details of techniques of measuring double stars, see Argyle, 2004. We must hope that the periastron passage is in the spring of 2005 and that the star does not move into conjunction before the most interesting part of the orbit is reached. Try to see it – the next periastron won't be until 2174!

## ACKNOWLEDGEMENTS

I am indebted to Christopher Taylor for drawing my attention to the detail of the observations made at Dorpat in 1836. I am also grateful to Mark Hurn, Librarian at the Institute of Astronomy, Cambridge, Mme Florence Greffe, Conservateur du Patrimoine historique at the Académie des Sciences in Paris, Dr Peeter Traat and Dr Izold Pustōlnik at Tartu Observatory, and Dr Brian Mason at the United States Naval Observatory.

## FURTHER READING

Argyle, Bob (ed.) (2004), *Observing and Measuring Visual Double Stars*, Springer.

Flamsteed, J. (1725), *Historia Coelestis Britannica*, Vol. 2, London.

Rigaud, S.P. (1832), *Miscellaneous Works and Correspondence of the Rev. James Bradley, D.D., F.R.S.*, Oxford University Press.

Lewis, T. (1906), *Measures of the double stars contained in the Mensurae Micrometricae of F.G.W. Struve*, Royal Astronomical Society.

Lewis, T. (1908), 'Double-Star Astronomy', *Observatory* 31, 88.

Smyth, W.H. (1860), *The Cycle of Celestial Objects continued at the Hartwell Observatory to 1859 (Speculum Hartwellianum)*, John Bowyer Nichols & Sons.

Struve, F.G.W. (1837), *Mensurae Micrometricae*, Petropoli.

# Titan: Lifting the Veils on a World of Intrigue

GARRY E. HUNT

A major highlight of the Cassini–Huygens mission to the Saturnian system will occur in January 2005, when a probe (named Huygens) will descend through the atmosphere of the planet's largest satellite. This will be the first probe encounter with Titan, the giant satellite of Saturn discovered by Christiaan Huygens in 1655. But why is this moon of Saturn so interesting and apparently important in our quest to understand the origin and development of the Solar System?

Titan is a planet-sized satellite, with a diameter of 5,150 kilometres (3,200 miles); it is second only to Jupiter's Ganymede in size as a satellite. It has a mean density of 1.88 g cm$^{-3}$ and therefore nearly twice the density of water, suggesting it is probably composed of half rock and half water ice. From its discovery more than 350 years ago, the subsequent telescopic observations from Earth were unable to resolve the disk of Titan. Only basic astronomical observations were made, and our knowledge of this distant body was very limited. A dramatic leap forward came in the early 1980s when the Voyager spacecraft swept past Titan (Figure 1). They discovered a nitrogen-rich atmosphere, which in many ways (apart from the fact that Titan is far colder) appears to resemble a very young Earth before the appearance of life, when oxygen probably existed in the form of carbon dioxide and supplies of methane were abundant. The clouds completely veil the surface of Titan, which could be covered with organic deposits falling like rain. Exploring Titan provides a chance to turn back the pages of time in a dramatic manner, and the Huygens probe provides the prospect of lifting Titan's veil for the very first time.

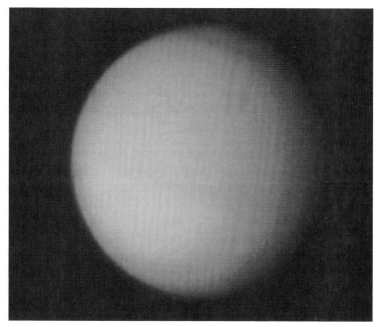

**Figure 1.** Titan shows little more than its upper cloud layers in this Voyager 1 picture, taken on 4 November 1980 at a range of 12 million kilometres (7.5 million miles). The orange-coloured haze, believed to be composed of photochemically produced hydrocarbons, hides Titan's solid surface from the Voyager cameras. Some subtle shadings in the clouds are visible, and Titan's southern (lower) hemisphere is brighter than the northern. (Image courtesy of NASA Jet Propulsion Laboratory.)

## TITAN'S ATMOSPHERE AND SURROUNDINGS

Titan's atmosphere is made up almost entirely of nitrogen with less than 1 per cent of methane in the upper layers. The surface pressure is 1.6 times that of the Earth, and with a surface temperature of a mere 92 K (about −180°C) considerably colder than our planet. But Titan and Earth have some similarities, since they are both bodies with nitrogen-rich atmospheres, so there is the possibility of learning more about the evolution and development of the early Earth, as well as planetary atmospheres in general, from this encounter with Titan.

Visually, some people may consider Titan to be a disappointment, since the surface cannot be seen at visible wavelengths. The Voyager observations in 1980 found the Titan surface covered in smog, formed from products of the complex methane and ammonia chemistry, which produces such gases as acetylene, ethylene, hydrogen, methyl acetylene and propane, all of which have already been found in the atmosphere. Carbon monoxide and carbon dioxide have also been discovered and, more recently, water vapour too.

In 1997, observations from ESA's artificial satellite ISO, the Infrared Space Observatory, in geostationary orbit around the Earth since 1995, provided evidence of water vapour in Titan's atmosphere. This may help to provide an explanation for the existence of oxygen compounds on Titan. The water vapour may enter Titan's atmosphere in the form of water ice particles sputtered from the nearby Saturnian rings and satellites, which then vaporize and combine with hydrocarbons producing the carbon monoxide and carbon dioxide we see today.

On Titan it is thought that about 95 per cent of the disassociated methane is irreversibly converted into acetylene and ethane. The constant removal of methane from the upper atmosphere by chemical reactions will require constant replenishment from the surface. The structure of the atmosphere will regulate these processes and the supply of materials from the lower to the upper levels in the atmosphere.

The atoms and molecules of hydrogen produced by photochemical reactions can easily escape from Titan as a consequence of its weak gravitational pull. However, they cannot escape from the pull of Saturn itself and the result is a doughnut-shaped torus of hydrogen that remains around Titan's orbit. This thick disc of hydrogen atoms surrounds Saturn and extends from the orbit of Rhea to just beyond Titan, giving a width of about 1,000,000 kilometres (over 600,000 miles). The torus co-rotates with the magnetosphere of Saturn. There are no observations yet to suggest that Titan has a magnetic field. If it does, however small, there will be some fascinating electrical currents and plasma interactions in this rotating torus.

Although molecules of hydrogen cyanide (HCN) have been found in interstellar space, their detection in the atmosphere of Titan was the first within an atmosphere in our Solar System. This discovery made from Voyager observations is very significant, since HCN is a key intermediate link between the synthesis of amino acids and the bases present in nucleic acids. It would be too simple to follow the argument towards Titan supporting life in any direct manner, especially at these very low temperatures. However, the presence of HCN is important in developing an understanding of atmospheric evolution.

But the atmospheric problems with Titan do not stop there. Titan's disk appears reddish or at least orange in colour, but the two hemispheres, viewed from above the cloud (smog) tops are not the same! At the time of the Voyager encounters, the Southern Hemisphere was uniform in brightness while the Northern Hemisphere was darker and redder. At this time the North Pole was covered by what appeared to be a dark hood, and then nine months later, at the time of the second fly-by, it was just a dark collar. Is this an example of high-level weather systems in Titan's atmosphere?

## THE SURFACE OF TITAN

We do not know much yet about the surface of Titan. There have been some tantalizing observations, and an increasing number of theories too, in recent years. But we shall have to wait for the Huygens probe observations for any first-hand measurements, as it lies hidden beneath thick layers of methane clouds and smog.

For a long time, it was thought that with the possibility of organic material falling on to the surface, and with ethane being the major product in liquid form, Titan's surface might be covered entirely with ethane–methane oceans as recent radio observations suggest. Certainly the surface temperature is close to the triple point of methane, which means that, like water on the Earth, it can occur simultaneously in the three states of solid, liquid and vapour. There could be methane rain and snow, possible cliffs of methane too, and even rivers and oceans, which in the colder locations would be rather slushy. But with a surface temperature of about 92 K, which varies by about three degrees between pole and equator, according to the Voyager observations, we anticipate some variations in the surface features of this strange body.

The observations made with the Hubble Space Telescope (Figure 2) may be the first clue, as they have provided the most interesting information to date concerning the surface of this distant body. Peter H. Smith of the University of Arizona Lunar and Planetary Laboratory and his team took the images with the Hubble Space Telescope during fourteen observing runs between 4 and 18 October 1994, when for

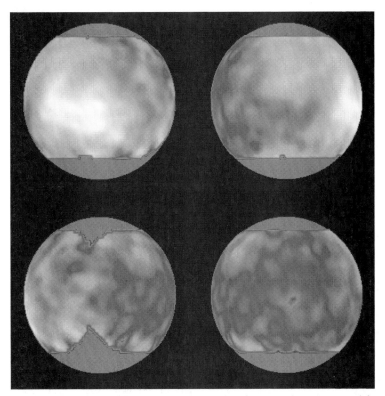

**Figure 2.** The Hubble Space Telescope glimpses Titan's surface in October 1994. Upper left: Titan's leading hemisphere (faces Earth at greatest eastern elongation). Upper right: the anti-Saturn hemisphere. Lower right: the trailing hemisphere (faces Earth at greatest western elongation). Lower left: the Saturn-facing hemisphere. Ground-based astronomers have established that, at longer wavelengths to which Titan's haze is nearly invisible, Titan is consistently brightest just after eastern elongation (between upper left and upper right images). The darkest side (from the ground based observers) is near western elongation (lower right). (Images courtesy of Peter H. Smith and the Lunar and Planetary Laboratory, University of Arizona.)

the first time they acquired images of the surface of Saturn's giant, haze-shrouded moon, Titan.

Smith and his team mapped light and dark features over the surface of the satellite during nearly a complete sixteen-day rotation. One prominent bright area they discovered is a surface feature 4,000 kilometres (2,500 miles) across, about the size of the continent of Australia. It is too soon to conclude whether these dark and bright areas in the Hubble Space Telescope images are continents, oceans, impact craters or other features. Scientists had long suspected that Titan's surface was covered with a global ethane–methane ocean. These Hubble images showed that there is at least some solid surface.

From spectroscopic measurements over almost ten years, we have discovered that the surface of Titan is not the same when you look at different longitudes or latitudes. The leading hemisphere is brighter than the trailing one. Certainly we should like to know if Titan's surface is mostly liquid, or if it is mostly solid, with ice and rocks sharing the landscape (Figure 3). The recent observations from the Arecibo

**Figure 3.** Artist's visualization of the surface of Saturn's largest moon, Titan, where the temperature is a frigid −180°C. Titan's atmosphere is far denser than Earth's and is primarily nitrogen laced with hydrocarbons such as methane, ethane, acetylene and propane. There may be deposits of black tar-like substances on the surface. Here, see a meandering ethane river as it flows around a line of rocky bluffs and into a lake. Pockets of methane fog drift across the lowlands. Here and there, large blocks of icy bedrock lie exposed on the surface. (Image courtesy of Bill Nuttall and York Films of England.)

Radio Telescope in Puerto Rico, suggesting the first evidence of liquid hydrocarbon lakes on the surface, enhance these views. The mystery of what is actually beneath the clouds of Titan remains unresolved – for the moment.

Although there may be some form of solid surface, the absence of an intrinsic magnetic field means that there is no electrically conducing core. Perhaps this suggests that there is a soft ice mantle beneath the crust and a relatively large silicate core. Any possible similarity between Titan and Ganymede will be particularly significant.

## WEATHER ON TITAN

Titan is yet another body that exhibits weather and climate variations where scientists can test their understanding of atmospheric processes and forecasting in regimes quite different from those found on the Earth. So far, no small-scale structures have been found in the atmosphere, although they are anticipated when higher spatial and temporal resolution observations are expected to be made during the Cassini mission.

Observations by Griffith, Hall and Geballe (*Science*, 20 October 2000) suggested that Titan's atmosphere possesses a hydrological cycle similar to the one on Earth with clouds, rain and seas, but with methane playing the terrestrial role of water. Surprisingly, their observations displayed enhanced fluxes of 14–200 per cent on two nights at precisely the wavelengths that probe Titan's lower-altitude clouds. The morphology of these enhancements at the four separate wavelengths observed suggested that clouds covered ~6–9 per cent of Titan's surface and existed at ~15 kilometres (9 miles) altitude. We could have wet weather on Titan too, but with methane and not water! Bright patches of methane observed in the area of the South Pole of Titan seen by Bouchez et al. (*Nature*, 19 December 2002) are thought to be cloud systems, and possibly even thunderstorms (Figure 4).

In the upper atmosphere, where there are 20 K temperature contrasts, there may be winds as strong as 100 ms$^{-2}$ (225 miles per hour). Indeed, Titan could be like Venus, possessing an upper atmosphere that rotates more rapidly than the lower levels. So far there is little more we can state with any confidence. The images showing hemispheric differences in the cloud contrasts would also suggest

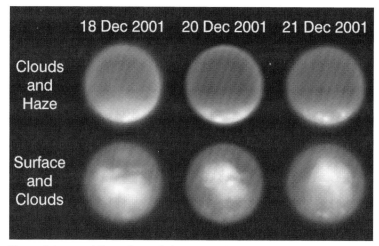

**Figure 4.** Images of Titan taken with the W.M. Keck II telescope during three nights in December 2001. The upper row shows images of just Titan's troposphere (lower atmosphere), which contains, in addition to the newly discovered south polar methane clouds, a 'haze' covering the south polar cap. The limb-brightening is also caused by a tenuous global atmospheric haze layer. Images in the lower row show Titan's surface rotating as well as the same methane cloud features near the south pole. (Images courtesy of H.G. Roe, I. de Pater, B.A. Macintosh, C.P. McKay and W.M. Keck Observatory.)

some inter-hemispheric circulations, but so far the precise details are unknown.

## PROSPECTS FROM THE CASSINI–HUYGENS MISSION

The Cassini–Huygens spacecraft has successfully entered orbit around Saturn. At 2112 PDT (Pacific Daylight Time) on 30 June 2004, the flight controllers had confirmation of the firing of the main spacecraft engine, which reduced its speed, allowing the craft to be captured by Saturn's gravity and enter a closed orbit. This begins a four-year tour of the ringed planet, its mysterious moons, the stunning rings, and its complex magnetic environment. During the Saturn tour, Cassini will complete seventy-four orbits of the ringed planet, make forty-four close fly-bys of the mysterious moon Titan, and numerous fly-bys of Saturn's other icy moons. If Cassini enjoys the same success as the

Galileo mission, we are likely to have a much longer mission than currently planned and even more observations.

The Huygens probe's descent into Titan's atmosphere in January 2005 must rate as one of the most exciting planetary encounters of all time. As we have seen, our current knowledge of this planet-sized satellite is intriguing and could provide some very important information associated with the development and evolution of planetary atmospheres in our Solar System and in particular in relation to the early Earth.

While the Cassini orbiter has been very busy and active on its way to the Saturnian system, the Huygens probe has remained relatively dormant throughout the 6.7-year interplanetary cruise, except for biannual health checks. Prior to the probe's separation from the orbiter, a final health check will be performed. The spacecraft will be loaded with the precise time necessary to turn on the probe systems, fifteen minutes before the encounter with Titan's atmosphere, and then the probe will separate from the orbiter and coast to Titan for twenty-two days with no systems active except for its wake-up timer.

The main mission phase will be the parachute descent through Titan's atmosphere. The batteries and all other resources are designed for a Huygens mission duration of 153 minutes, corresponding to a maximum descent time of 2.5 hours plus at least three additional minutes, and possibly half an hour or more, on Titan's surface. The probe's radio link will be activated early in the descent phase, and the Cassini orbiter will 'listen' to the probe for the next three hours, which includes the descent time plus thirty minutes after impact. Not long after the end of this three-hour communication window, Cassini's high-gain antenna (HGA) will be turned away from Titan and towards Earth.

During its descent, Huygens's camera will capture more than 1,100 images, while other instruments will directly sample Titan's atmosphere and determine its composition. As it finally enters Titan's atmosphere, three sets of parachutes will slow down the probe and provide a stable platform for scientific measurements. The fully instrumented robotic laboratory will reach the enigmatic Titan's surface about two and a half hours later.

There are six instruments on board the European Space Agency probe, Huygens, designed to address the major science objectives of this part of the mission, namely:

- To determine the most abundant elements, and establish the most likely approach for the formation and evolution of Titan and its atmosphere.
- To determine the relative amounts of different components of the atmosphere.
- To observe the vertical and horizontal distributions of trace gases; search for complex molecules; investigate energy sources for atmospheric chemistry; determine the effects of sunlight on chemicals in the stratosphere; study formation and composition of aerosols.
- To measure winds and global temperatures; investigate cloud physics, general circulation and seasonal effects in Titan's atmosphere; search for lightning.
- To determine the physical state, topography and composition of Titan's surface; characterize its internal structure.
- To investigate Titan's upper atmosphere, its ionization and its role as a source of neutral and ionized material for the magnetosphere of Saturn.
- To determine whether Titan's surface is liquid or solid, and analyse the evidence for a bright continent as indicated in Hubble Space Telescope images taken in 1994.

To achieve any of these objectives will provide a significant advancement in our basic knowledge. We can expect there to be many exciting and unexpected measurements and surprises too, which will give us new insight into the formation and evolution of Titan, the Earth, and hopefully planetary atmospheres in general.

# Winds from the Stars

CHRIS KITCHIN

## INTRODUCTION

We are all familiar with winds, from the gentle zephyrs of a summer's evening to the raging tempests of hurricanes and the even more damaging blasts within tornados. Space looks calm in comparison with the turbulent Earth, yet that is deceptive. There are winds in space that make the worst cyclone seem a mere trifle.

Terrestrial winds arise from pressure differences within our atmosphere. Those pressure differences in turn are caused by differential heating effects. Many of the winds in space likewise occur through the heating effects of stars, and others develop through the alternative ways in which stellar energy can affect material around the stars.

For many years it has been evident that material flows out from stars. Nova and supernova explosions clearly blast material outwards at enormous velocities, though for only brief intervals of time. Stellar winds are a more continuous emission of material. Thus steadier, less violent and longer-term outflows than those of novae must have produced the planetary nebulae that surround many old stars (Figure 1 and the front cover illustration of this *Yearbook*). There is also evidence of outflows from stars' spectra. Occasionally a star's spectrum will show emission at the normal wavelengths of spectrum lines, together with absorption on the short-wavelength edges of the emission (Figure 2). This shape arises when a hot gas flows away from the star and accelerates as it moves outwards. The emission component is produced in the hot, slow-moving inner regions of the outflow, and the absorption component by the cooler, faster-moving, outer parts. Line shapes of this type are known as P Cygni profiles, after the star P Cygni in whose spectrum they were first noticed.

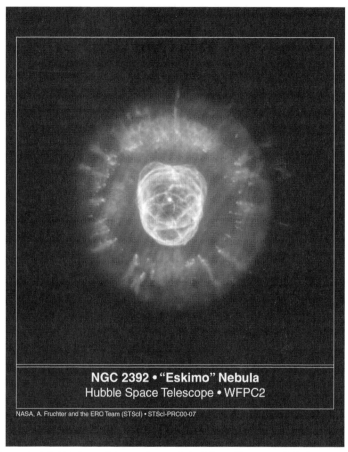

**NGC 2392 • "Eskimo" Nebula**
Hubble Space Telescope • WFPC2

NASA, A. Fruchter and the ERO Team (STScI) • STScI-PRC00-07

**Figure 1.** A Hubble Space Telescope image of the Eskimo Planetary Nebula (NGC 2392). The material surrounding the central star has originated from a stellar wind blowing outwards at about 30 km/s for some 10,000 years. A subsequently developing faster wind expanding at 400 km/s and interacting with the slower-moving material has then produced the complex shapes observed. (Courtesy of NASA/STScI, Andrew Fruchter and the ERO Team [Sylvia Baggett, STScI, Richard Hook, ST-ECF, Zoltan Levay, STScI].)

Violet                                                                    Red

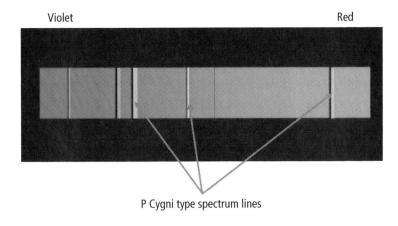

P Cygni type spectrum lines

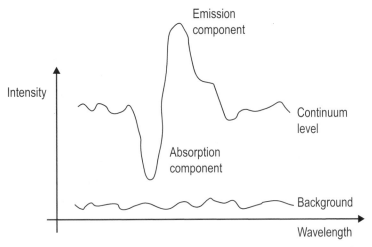

**Figure 2.** A stellar spectrum with P Cygni type lines (top) and a tracing through one of the lines (bottom).

## THE SOLAR WIND

The suspicion that such winds might occur much closer to the Earth did not arise until the 1950s. Ludwig Biermann (1907–86) noticed that radiation pressure was sufficient to push the dust tail of a comet away from the Sun, but was not adequate to explain the gas or ion tail

(Figure 3). He therefore suggested that a stream of particles, the solar wind, must affect the ions. In 1958, Eugene Parker (1927–) predicted how such a flow of particles might originate from the Sun, and in 1959 the Luna 2 spacecraft detected the solar wind directly.

Four decades later, and because of observations from many spacecraft (Figure 4), we now know the properties of the solar wind in detail. Puzzles remain, however, about exactly how it originates. The solar

**Figure 3.** Comet Hale-Bopp in March 1997 showing the ion tail produced by the solar wind (on the left and straight) and the dust tail produced by radiation pressure (curving to the right). (Courtesy of C.R. Kitchin, 1997.)

**Figure 4.** An artist's impression of the SOlar and Heliospheric Observatory (SOHO). The spacecraft has observed many aspects of the Sun, including measuring details of the solar wind, with great success ever since its launch in 1995. (Image courtesy of SOHO consortium. SOHO is a project of international cooperation between ESA and NASA.)

wind is formed mostly from protons and electrons together with a few helium nuclei and the occasional ion of a heavier atom. Near the Earth the density is about 5 to 10 million particles per cubic metre. This may sound a lot, but by terrestrial standards it is a very hard vacuum – there are some 300 septillion (300 plus twenty-four zeros) molecules per cubic metre in the atmosphere at the surface of the Earth. The solar wind has two components: the slow and fast winds. The slow wind is confined to the Sun's equatorial regions around sunspot minimum, but has a wider distribution when the Sun is more active. At the Earth's distance from the Sun, the particles are travelling at around 350 kilometres per second, though this can vary markedly with solar activity, and the wind is very gusty. The fast wind emerges from solar latitudes above about ±20° (Figure 5). It is much less variable and the particle velocity is around 700 kilometres per second.

The Sun loses about one million tonnes of material per second through the solar wind, or about a quarter of the mass that is converted to energy each second in order to supply the Sun's energy. Though this appears to be a great deal of material, it is very tiny compared with the mass that is lost from some stars. Furthermore, on a solar scale it is not significant – the Sun will lose only 0.02 per cent of its total mass via the solar wind over the whole of its 10,000-million-year main sequence lifetime.

The cause of the fast component of the solar wind is the high temperature of the solar corona. This is about 1.5 million K at the base of the corona and decreases slowly on moving away from the Sun. At 25 solar radii (18 million kilometres) the temperature has dropped to about 600,000 K. Protons at that temperature move on average at about 120 kilometres per second. However, at 25 solar radii, the escape velocity from the Sun has dropped from about 600 kilometres per second at the photosphere to 110 kilometres per second. From that point outwards, the protons can therefore escape directly from the Sun into space. Electrons, of course, are moving forty times faster than

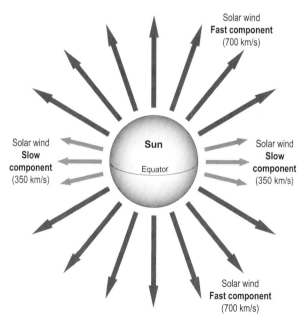

**Figure 5.** The slow and fast components of the solar wind at around the minimum of the sunspot cycle.

protons and so exceed the Sun's escape velocity even within the photosphere. The Sun's gravitational field is thus too weak to hold on to the corona, and it slowly leaks into interstellar space – the phenomenon that we call the solar wind.

The temperature (and hence the pressure) within the corona suffices to accelerate the particles up to about 200 kilometres per second. However, current models suggest that this is an upper limit. Thus in order for the particles to reach the 700 kilometres per second observed for the fast component of the solar wind, an additional accelerating mechanism is needed. It is not clear currently what form that additional accelerating mechanism takes. One suggestion is that the extra

**Figure 6.** Coronal streamers are associated with the slow component of the solar wind. Image from the SOHO spacecraft. (Image courtesy of SOHO/LASCO consortium. SOHO is a project of international cooperation between ESA and NASA.)

energy comes from plasma waves, which are waves resulting from the interaction of the ionized coronal gas and magnetic fields, but this suggestion is quite unproven.

The slow component of the solar wind seems to be associated with coronal streamers (Figure 6), but again the exact mechanism whereby its particles are accelerated is unclear.

## STELLAR WINDS

The loss by the Sun of a million tonnes of material per second corresponds to it losing $10^{-14}$ solar masses per year. Such a weak wind would not be detectable for a more distant star, although presumably other solar-type stars do have similar winds to that of the Sun. The stellar winds that we can observe are thus a great deal stronger than the solar wind. Mass loss rates from $10^{-9}$ to $10^{-3}$ solar masses per year are possible, and wind velocities can attain 3,000 kilometres per second. At the top end of this range the mass loss rates are so huge that large stars may halve their masses during their main sequence lifetimes. Observable stellar winds are found among three main groups of stars: the young, the large and the hot.

Young stars such as T Tauri stars often have P Cygni-type line profiles in their spectra. The expansion velocities obtained from these lie in the region of 100 to 200 kilometres per second, and represent mass loss rates of $10^{-9}$ to $10^{-7}$ solar masses per year or more. The wind is very variable in its strength, both between stars of similar types and for a single star over time. Often the outflow is in the form of two jets, with the material escaping along the poles of the star, while an equatorial disk suppresses the wind in that plane (Figure 7). There is still much to be understood about the outflow (and inflow) of material within young stars, but it seems likely that a coronal-type origin contributes to at least a part of the stellar wind.

Many cool supergiants and long-period variables show evidence of winds with velocities of 10 to 100 kilometres per second. Sometimes this can be detected from the P Cygni line profiles in the stars' spectra. However, the stellar wind lines often blend with those from the star, making their interpretation difficult. When the star is a member of a binary system, the spectrum lines from the wind can then sometimes be found in the spectrum of the companion. The mass loss rates are

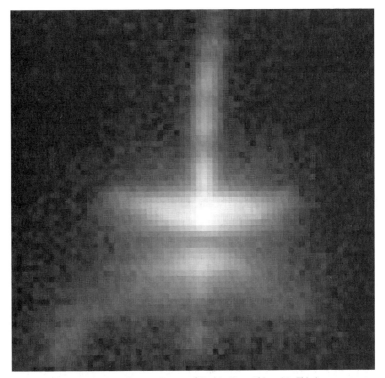

**Figure 7.** A Hubble space telescope image of Herbig-Haro Object 30. This is a young star losing mass via a stellar wind that is constrained by an equatorial disk into two jets. The star itself is hidden at the centre of the edge-on dark disk of material. (Courtesy of Chris Burrows, STScI, the WFPC2 Science Team and NASA.)

estimated to range from $10^{-9}$ to $10^{-5}$ solar masses per year. For example, Antares (Alpha Scorpii) is losing $10^{-6}$ solar masses per year at 17 kilometres per second. The winds in these stars, especially the cooler ones, may be driven by dust particles. Infrared emission from the particles shows that carbon and silicate grains exist near the stars. Radiation pressure may then drive the dust grains outwards. The collisions of the dust particles with atoms in the surrounding gas, and the viscosity of the gas, will result in the outward force being communicated to the whole of the outer layers of the star.

The most intense stellar winds, however, come from the hot stars. O-type stars, which have temperatures from 30,000 K to 50,000 K, can lose up to $10^{-7}$ solar masses per year. Wolf-Rayet stars have similar temperatures to the O-type stars, but different abundances of the elements, and their winds can reach velocities of 3,000 kilometres per second and mass loss rates over $10^{-4}$ solar masses per year. The central stars of planetary nebulae (Figure 8) are even hotter, perhaps up to 150,000 K, and their wind speeds can reach 10,000 kilometres per second. However, because these stars are small, the mass loss rates peak at perhaps to $10^{-5}$ solar masses per year.

If the winds from stars with mass loss rates hundreds of million times that of the Sun arose from the same thermal mechanism as the

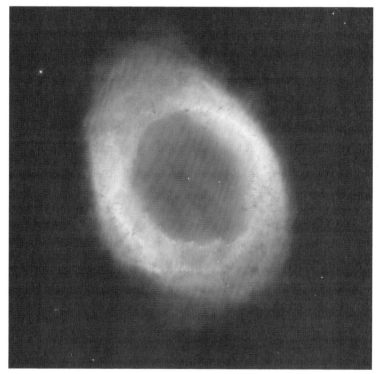

**Figure 8.** A Hubble Space Telescope image of the Ring Nebula (M57, NGC 6720). The material has been emitted as a low-velocity stellar wind to form a uniform sphere. A later, much higher-velocity wind has then swept up the inner parts of the sphere to leave a spherical shell, which we see in projection as a ring. (Courtesy of Hubble Heritage Team and AURA/NASA/STScI.)

solar wind, then the coronae would need temperatures of hundreds of millions of degrees. Such high temperatures can be ruled out, since they would produce other effects, such as X-ray emission, that we do not find. A different mechanism is thus needed for producing strong stellar winds.

That mechanism is radiation pressure. An atom or ion near a star can absorb and then re-emit radiation. The momentum of the absorbed photon is transferred to the atom, and then lost again when the photon is re-emitted. However, the absorbed photons all come from one direction – from the star – while the emitted photons can go in any direction (Figure 9). There is thus an outward force on atoms close to stars. That force can be several hundred times the force due to gravity, even for massive stars. Hence the atoms and ions can be driven outwards to form a stellar wind.

There is, nevertheless, a problem with these radiation-pressure-driven winds. The observed wind velocity is highly supersonic (the speed of sound in the material near a star is 50 to 100 kilometres per second). However, if the force on the material is too high initially,

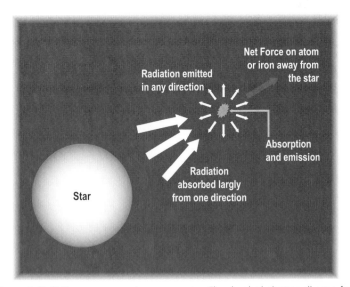

**Figure 9.** Radiation pressure on atoms near a star. The absorbed photons all come from one direction – from the star – while the emitted photons can go in any direction. There is thus an outward force on atoms close to stars.

it will not accelerate through the speed of sound, but instead become a subsonic and highly turbulent flow. The material therefore has to accelerate gently until it becomes supersonic. After that velocity, the accelerations can be limitless and thus lead to the intense winds that we observe near hot stars. However, it is not clear how that initial gentle acceleration can be produced.

There are two suggested mechanisms. The first is that there is a very thin coronal stellar wind close to the star that accelerates the material to beyond the speed of sound. The second is that the star itself has strong absorption lines at the same wavelengths that the outer material is absorbing. The force on the atoms is thus much reduced. Later, as the material is moving faster, Doppler shifts move the stellar absorption lines away from those of the accelerating material, and the latter is exposed to the full radiation intensity from the star. Either way, for hot stars, winds are produced whose speed and intensity will cause the star to lose so much mass that its evolution will be changed.

The winds from hot stars, and indeed all other stellar winds, make those on the Earth seem puny in the extreme – though whether that will be a comfort next time you have to replace roof tiles after a storm is doubtful!

# The Eclipse of a Lifetime

## MICHAEL MAUNDER

### SEIZING THE OPPORTUNITY

An old saying goes something like this: 'It's not the things you do you regret, rather those you don't.' So when I discovered only a single seat remained on a flight to see the November 2003 total solar eclipse over Antarctica, I booked immediately.

Some months later as the departure date approached, the uniqueness of the trip became clearer. We would get within 0.6 second of the maximum totality possible anywhere, with an added bonus that the apparent altitude of the Sun above the horizon would be at least 3° more than on the ground. Instead of about 1 minute 55 seconds maximum duration, we would get an extra 36 seconds or over 2 minutes 30 seconds and an apparent elevation of 18° at 35,000 feet (over 10,000 metres) against 15° on the ground. The timing was also very peculiar because the shallow shadow angle meant that an observer on the ground, viewing at the same instant as us, would be in a different time zone since these change very rapidly that close to the pole at near latitude 70°S. Moreover, totality occurred the best part of a day before the calendar date of just a few minutes' flight away. Timing for us was on 23 November 2003, 22h 44m Universal Time, with our local time based on Australia in the morning of 24 November, having, in effect, been up all night with take-off at 02.00 hrs on the 24th. All very confusing! With crystal-clear skies virtually guaranteed in the stratosphere, all that remained before leaving the UK was a final check on solar filters for safe viewing. Normally the concern is quite the opposite – that of seeing anything at all through clouds.

The specially chartered 747-400 flew from Melbourne, almost as close as one can get to Antarctica. Travel options were very simple – get there and back as quickly as possible because that same weekend coincided with the Rugby World Cup Final. Flights from the UK were limited, and I went by the fastest route via Singapore to arrive overnight. This was a wise choice because the Sun had been extremely active for some weeks, and while we flew across Australia some superb displays of the Aurora Australis were seen from the window. (As far as locations in the Southern Hemisphere are concerned, Melbourne is relatively close to the South Magnetic Pole.) A good start.

The next day required a final equipment check. One of the golden rules of eclipse chasing is to take more than you need because 'Murphy' lurks. Although eclipse chasers had reserved seating, only about fifty in the 'A' row, on the left-hand side of the aircraft, were suitable for effective eclipse viewing. Quite a few 'A' seats had no window at all and these, together with all those on the right-hand side of the plane, had been set aside for regular explorers wanting to view and photograph Antarctica. They were involved in a complex game of 'musical chairs', so that everyone got a good view part of the time. My particular problem was not knowing in advance exactly how the seats matched with the windows. Would it be possible to use a tripod or must one contort oneself for hand-holding? Being prepared for anything turned out to be a wise move. The day before the flight should have been a rest day: it wasn't, as the Rugby Final was compulsory viewing.

## ABOARD THE AIRCRAFT

The day of the flight was also an odd one, since take-off was scheduled after midnight, with check-in when one would normally be going to bed. Fortunately, nobody got the dates wrong and turned up the next day! My decision to arrive in Australia close to the date of the eclipse paid off, and I was still on UK 'morning coffee' time. There was plenty of time in the departure lounge to see the plane arrive from Sydney and greet friends from previous eclipses. Checking in had some very unusual features. Our boarding passes had already been posted to us, and as these were valuable souvenirs the electronic verifier had to be reprogrammed to stop it keeping bits of it. The departures board also gave our destination as Melbourne, which was our place of departure: a good photo opportunity. Never in all my experience has a 747 taken so long to take off. Although none of us had hold luggage, the eclipse specialists were allowed almost unlimited carry-on, but that didn't account for it, and we settled down to a long haul. The mystery was solved shortly afterwards: it was the huge amount of drink on board for celebrations later!

My seat 64A proved to be almost ideal for viewing, but impossible to operate a tripod in the narrow seat raking without knocking it with the slightest movement inevitable from seats B & C. These had been allocated to other eclipse enthusiasts who were prepared to take pot luck for the eclipse, but obviously keen to get as close as possible to the window. Everyone was free to walk about anywhere on the aircraft at all times, in itself an extremely unusual thing in these security-conscious days, but the non-eclipse viewers were not allowed in our area for the critical ten minutes either side of totality. Over Antarctica at above 10,000 metres the risk of severe air turbulence is minimal, and that allowed the unique social gatherings to continue throughout the flight. Australia was soon well behind us with enough time to run through the equipment and practise the best options.

Although Glenn Schneider, the tour leader, had thought of every-thing and the plane had amazingly clear windows, there was no getting around the inevitable major snag of photography from an aircraft – the multiple reflections and images seen from the many window surfaces. All cameras using autofocus, and most do so these days, cannot avoid homing in on the sharpest and therefore the wrong picture. The wider the lens angle chosen, the more likely this phenomenon becomes. Even with our absolutely clean windows, another ugly snag cropped up very soon – the formation of ice crystals within the double glazing. No autofocus camera is smart enough to select the right image a long way off when it can 'see' bright ice crystals within its field of view. Ice crystals formed very quickly because the weather over Antarctica was particularly bad for the time of year. Other eclipse viewers on board the Russian icebreaker below us, and close to the ice shelf, only saw their eclipse through the cloud cover. Nowhere on that part of the continent, with its many permanent ground bases, was it completely clear, and the pilot announced that a severe hurricane-force storm was raging below, unusual for the time of year. The further inland we flew, the worse the storm below became, and even at over 10,000 metres the plane could not avoid clouds. The cloud ceiling remained way up above, dashing any hope of doing much sightseeing before the eclipse.

As the plane got closer to continental centre, and the eclipse track, the cloud cover vanished magically and the plane emerged into bright sunlight, giving us our first glimpse of the 'White Continent' after many hours in the air: time for a little sightseeing and a better understanding of the vastness of the place. It was also a magical short interlude because the cloud cover had not completely cleared below and as we circled some superb 'glories' appeared. The phenomenon is caused by light reflected back from ice crystals in the clouds and is seen as a darker multi-coloured circle around a brighter patch of light containing a distinct shadow of our plane. The exceptionally clear air above us allowed full sunlight to show the effects in all its glory, hence, one assumes, the name. The delight didn't end there; as we circled the display got larger and smaller depending how far up we were from that particular bank of cirrus. At one stage we passed through a layer and the effect was startling with the glory cut in half, the upper portion being considerably fainter than that below, against a perfect back-drop of clean white snow. It was a unique experience and a real bonus to the trip.

Emerging from the cloud with its high humidity worked wonders with the ice crystals in my window, and before long nearly all had gone, but as I went around the plane for a final social chat many photo-graphers had to contend with this stuff all the way through. Anyone intending to do serious photography from a plane window must always be prepared to see nothing but a frosted window – very much a lottery. Then there is another major hazard to expect. Even if a photographer disregards the very real risk of either damaging one's eyesight or the camera 'innards' shooting into the Sun, and astronomers will always know which solar filters are essential and will have them, the contrast is poor to abysmal. Polarizing filters help to some extent, but never allow really good snow-scene pictures with the Sun more or less straight out of the window, almost straight into the lens. Getting there with the Sun in just that position was the whole point of the flight plan! Polarizing filters are virtually ineffective when used in conjunction with high-density solar filters.

## FIRST CONTACT

First contact occurred exactly on schedule and was well seen by the serious eclipse photographers who, by and large, did not do much through cameras. The plane was still doing circles to get in as much sightseeing as possible and a stable image became very difficult. I found straightforward naked-eye observation through my solar filter a lot more interesting, and also a lot easier to handle with the micro-turbulence on the aircraft impossible to overcome without a gyro or similar aid. At this very late stage, I decided to jettison any idea of photographing the eclipse totality with a conventional camera and telephoto lens. 'Murphy' would strike and a minor judder at the critical moment of exposure would lose any unique pictures. Modern stabilized-image lenses could work very well, although beyond the budget of many. Better to stick with wide-angle lenses, and also video, where the eye would accommodate the tremors in the replay. That decision paid off handsomely.

Without the tripod for stability one of the oldest tricks in the photographer's armoury really worked well – the 'bean bag'. Normally one buys this or makes one from a bag filled with any old granular material like rice grains. That would mean extra weight taken on board and, to me, the obvious substitute was already there and free, the airline pillows. As a tip for future reference, try this one. Whenever you don't have a stabilized lens and want to stop most of the plane vibration and your hand's unsteadiness, simply put the pillow on to the window ledge, then rest as much of the camera on it as you can without actually touching the lens against the window. The way this stabilizes the image is nothing short of miraculous; it costs nothing and there is no tripod to trip over.

As totality approached, the excitement increased and the plane set on to the pre-determined course and altitude. The Sun stayed rock steady at 90° out of the window. Final checks were made on the equipment. During the eclipse's partial phases we got many opportunities to see and note the huge sunspots being eclipsed in turn. It is rare to see so many and that large in area. By the time the eclipse was about three-quarters into the partial phase, very little drop in light level was apparent to the eye because the brain can accommodate this. Then the peculiar ashen hue became pronounced. With only a few minutes to go, the eclipse viewers were left alone and that side of the plane was proclaimed 'out of bounds' to anyone else. Some quite amazing colours appeared on the ice shelf below as the Sun became more obscured, ranging from deep yellows through oranges to a purple hue. The pure white of a few minutes before had vanished. Similar colours were also seen on the horizon, with much stronger reds seen at the Earth's curvature.

With about five minutes to go, the coloured shadow on the ground became the most amazing feature as the plane raced to keep ahead of it. The plane was flying at about one-quarter of the speed of the approaching shadow, and that is why we got over an extra half-minute of totality by appearing to slow down the actual speed of the shadow across the ground. The shadow showed up as a dark band right behind us, with the spectacular bands of colour merging into near darkness. Even though we still hadn't got into totality, the sight of that pursuing shadow was well worth the trip on its own, but the sight of the chasing 'brightening' after eclipse was even more memorable. Venus had been a clearly visible sight for quite a long time. As the sky took on a dark purple tinge, suddenly you became aware that it was there.

I had made up a special mount for the video camera so that it was possible to hold it without needing to touch the camera to remove the solar filter. The filter slipped into a slot and as totality approached it needed no effort to take this away completely and place it where it could be retrieved immediately after totality. Experience had shown that for about fifteen seconds before true totality, it was possible to shoot without damaging the sensor and electronics. Readers should be aware that this does not apply to many cameras, and the filter should only be removed a second or so before totality. This always leads to additional stress and consequent camera shake. The very sensitive nature of the detecting system also meant that while one could see Baily's Beads, these were completely burnt out on the minute video image, as was the chromosphere. To get that would mean a much higher lens magnification, impossible in my circumstances.

On my video recording the aircraft noise was impossible to eliminate although I had all manner of microphones with me to try out. The noise, as always on aircraft, is all-pervading. However, Glenn gave a countdown and it is clear from the commentary that modern CCDs are indeed so sensitive that they record the bright inner corona and chromosphere a few seconds into true totality, and similarly shorter at the end. The eclipse appears too short on video but was obviously clear-cut to the eye.

To the eye? Well, that is another trick I had learned over the years from microscopy, and then telescopy. With training, it is possible to look through the video viewing eyepiece with one eye and keep it centred, and still use the other eye to see and scan the scene in front of you in all its glory. In that way, I was able to see the greatly magnified image in the viewfinder, and soak up the unique spectacle out of the window with the other; this time a quite mind-blowing experience. That is how I was able to watch the shadow phenomena and anticipate when to remove and replace the filters, while still keeping my options open via the broadcast commentary from the flight deck.

## TOTALITY!

As the last vestiges of sunlight were blocked out by the Moon, Mercury became clearly visible to join Venus as two bright spots of light just over the plane's wing. At one time I had been concerned that the wing with the spoiler at the end would intrude on the eclipse and interfere with autofocus, but in fact it enhanced the visual imagery in a most artistic way (Figure 1). It was unnecessary to switch to manual focus because the solar corona almost filled the screen, and that focus still held when I tried zooming to a wide-angle shot to match what I was

**Figure 1.** When aligning his camera to photograph the eclipsed Sun, the author had been concerned that the plane's wing with the spoiler at the end would intrude on the eclipse and interfere with autofocus, but in fact it enhanced the visual imagery in a most artistic way. (Photograph by the author.)

seeing 'for real' with the other eye. The most striking feature of the corona was its sheer brilliance, much brighter than I had seen from the Chilean Altiplano a decade earlier. A high bank of thin cloud had taken the edge off on that occasion, whereas at our high altitude and no atmospheric pollution the full beauty could be appreciated. The coronal shape was also very similar to that earlier experience at a similar phase in the sunspot cycle after solar maximum: one very long streamer or 'beard' to one side and a pair of 'horns' on the other. This time the image was twisted to the left so that the appearance of a devil's face was not as pronounced as it had been in Chile. With the naked eye, no prominences could be discerned, although some were recorded by others using higher magnifications.

Now to the tricky bit. Because of my decision not to use telephoto lenses, I could devote my whole attention to leaving the video running the entire time and enjoying the view, perhaps the real point of the trip. Every eclipse chaser should adopt the same philosophy: 'If all the "techie" stuff fails, just lie back and enjoy the view.' That mental image will always stay with you, and there is nothing worse than an alternative memory of just the failed equipment, and control buttons that don't work. The tricky bit was how to get a picture of the general scene out of the window. Zooming out to a wide-angle on the video was not as easy as I had expected, and for the usual reason: getting a level horizon. Somehow it is always easier to line up the horizon when the image is magnified, and it took precious seconds canting the video while operating the zoom control. In the end I got it right. Then I got greedy. Why not try for an ordinary camera wide-angle that I had set up ready?

That is the occasion when a schizoid mentality dealing with two different eyes' images helps. Keeping the video on target was the easy bit, but picking up a separate camera and hand-holding it in correct alignment without looking through the viewfinder does take a lot of practice. Then you need a steady hand when highly tensed by the emotion of the spectacle in front of you! An activity such as rifle shooting has a lot to commend it! That, and no pre-eclipse drinks. Most of my camera shots were sharp. Even with my very steady hand, they certainly would not have been had I used the slightly higher focal length standard lens.

Experience certainly teaches a few lessons and another I find invaluable is this. 'When in doubt, never try for telephoto pictures. Go for wide angles.' Somehow the visual impact of the colours seen at the

**Figure 2.** With the plane set on to the pre-determined course and altitude, the Sun stayed rock steady at 90° out of the window. The plane had amazingly clear windows, with little evidence here of the multiple reflections and images seen from the many window surfaces – an inevitable snag of photography from an aircraft. (Photograph by the author.)

picture's edge where the lunar shadow occurs more than makes up for the missing high detail. On every eclipse trip there is always someone with more effective equipment than you will ever aspire to, and their pictures can be added to your more personal ones. There is nothing worse than a blurred or distorted corona through being too ambitious.

The flight deck announced a warning when the total phase was about to end, but with my eye on the chasing end of the shadow on the ground there was plenty of time to soak up the view and really enjoy it. As third contact approached, the video imagery brightened very noticeably, another useful early-warning feature. The glorious spectacle ended with the most beautiful 'Diamond Ring' ever, and it was necessary to be a lot slicker in getting the safety filter back on. Unlike second contact, where dangerous light levels are decreasing, here we are dealing with a huge build-up and within a few seconds the CCDs fully overloaded and the experience was over until the next one in a few years' time. There was only one other view to record properly and that was Venus still visible for several minutes after totality. The wing I had not been too keen on earlier was now a positive help, and with some care it screened the Sun from directly shining into the lens. Venus remained very obvious long after the plane banked and started its journey home.

## POST-ECLIPSE CELEBRATIONS

The non-eclipse passengers probably had a real bargain from that time on, as they had the major part of the partial phase to themselves, while we had to make do with a mad rush to the bar to celebrate our wonderful experience. The reason why the plane had needed such a long take-off run revealed itself. The bar was choc-a-block with Australian champagne – the heavy bottles weighed more than the drink inside. There were, it seemed, unlimited supplies although we did our best to drink it all. Time also to swap tales and experiences with others on board, and as I moved about there was a distinct 'coolness' on two counts. Many windows were heavily iced up and I wondered how many pictures would have come out. As the sun became more eclipsed, the solar heating must have dropped and allowed the vaporized water to recondense, then freeze: something to be aware of on any future eclipse flights. The other coolness was in my choice of flag. Many eclipse chasers proudly display their national flag and I had done the same. This was from Alderney in the British Channel Islands. It is the same as the English one but with a single lion at the centre, a distinction few would notice. As we were only a day or so from the English Rugby triumph, the reception it received on a Qantas plane is hardly surprising. More drinks to celebrate that.

Even with the long flight, and adrenalin draining away, sleep was not an option. I had already been awake more than thirty hours, yet it was still not midnight back in UK timing. The dramatic Antarctic scenery and the chance to chat and swap stories proved irresistible, and a chance to ponder on what to do next to top the experience. By the time we approached Melbourne again, many of my companions had slept for much of the time and I wondered why they were not interested in the later celebrations and unique views. Some had travelled even further to get on the trip.

Landing was apparently uneventful; then the pilot made a surprise announcement. We knew that the trip was a unique first as an eclipse seen from Antarctica, and that was still true. What Captain John Dennis then told us was that we had gone into the record books for the longest ever domestic flight! Although we had been to Antarctica, the plane had never landed and had returned to the same country (same airport, in fact) and that classed it as a domestic flight. It was a minute or two under fourteen hours. Maybe that explained the number sleeping. Fond farewells to all our new friends and then back to the hotel for a well-earned rest.

The next day's sightseeing in Melbourne ended with another rare sight for Northern Hemisphere astronomers. Venus, then Mercury, in a line the wrong way round with a thin lunar crescent pointing to the left and a glorious sunset over Melbourne Airport. The exceptionally clear air displayed a very bright earthshine effect, made even more dramatic by aircraft taking off and transiting the celestial grandeur. Many regrets that I could not follow the cycle over a few days, as the next day I boarded a return flight home, making a round trip of eight days. Memories there were in abundance, and there was no doubt at all that an experience like that will never happen again. It all goes to prove that, to get the best out of any similar trip, practising and knowing all there is to know about one's equipment, and taking back-ups, will pay handsome dividends.

# RAVEing Mad

FRED WATSON

If you have ever studied English at school or college, you might be familiar with a little conundrum designed to set you thinking about the subtleties of grammar. It comes straight from pedants' corner, and it goes like this. Turn the following into a correctly punctuated sentence:

> Smith where Jones had had had had had had had had had had had the approval of the examiners.

It seems impossible that such a bizarre string of 'hads' could ever make sense, but once they are tagged with the right punctuation, the meaning becomes perfectly clear. Doesn't it?

Now switch from English to astronomy, and imagine yourself looking out into a night sky radiant with stars. Seen at face value, they are all the same. Some are brighter than others, of course, and there are subtle colour variations here and there, but apart from that, the only thing that differentiates one from another is its position in the sky. And if we did not have the good old join-the-dots constellations, even that would be tricky without an instrument of some kind.

The message? Well, stars are not much different from unpunctuated 'hads' when it comes to trying to make sense of them. Little wonder the ancient watchers of the skies thought the stars were all stuck to the inside of a sphere, never changing or moving with respect to one another. In fact, one of the first big challenges of telescopic astronomy was to understand the three-dimensional distribution of stars in space. It was William Herschel who made the big breakthrough towards the end of the eighteenth century. He recognized that we live in a flattened disk of stars, though it was to be well over a hundred years before our modern picture of the Milky Way Galaxy emerged.

During the nineteenth century, astronomers realized they could detect the individual motions of stars, initially across the line of sight (known as 'proper motion') and later towards or away from us (the so-called 'radial velocity'). Whereas proper motions are derived from accurate positional measurements over a period of time, radial velocities are measured by forming the rainbow spectrum of a star and measuring the minute shifts of its features resulting from the Doppler effect. The spectroscope thus becomes a kind of star speedometer. Gradually, as the nineteenth century progressed, the 'punctuation' of the stars was being put into place.

When both the radial velocity and proper motion of a star are known, its true velocity through space (relative to the Sun) can be determined. Combined with a knowledge of its distance (usually obtained from brightness measurements in several different colours), this gives all the vital statistics on the position and motion of a given star within our Galaxy. For any one star, that is not particularly exciting, but when these data can be collected for thousands of stars, they start to reveal patterns and show up bulk motions within the Milky Way Galaxy. And that is much more interesting.

The most basic of these bulk motions is what is known as star-streaming, an effect discovered in the early twentieth century by the Dutch astronomer Jacobus Kapteyn (1851–1922). The mystery of this preferential motion of stars in particular directions was brilliantly solved in the 1920s by Bertil Lindblad (1895–1965) and Jan Oort (1900–92), who showed that it is simply owing to the way our Galaxy rotates. Stars whose orbits are closer to the galactic centre than the Sun will tend to overtake the Sun on the inside lane, whereas those with more distant orbits will seem to lag behind.

But what might you find if, instead of analysing the motions of thousands of stars in the Sun's neighbourhood, you could look at hundreds of thousands – or even millions? Such possibilities were not open to Oort and Lindblad, but they are to us today. And they promise to reveal *much* more than just the rotation of the Galaxy.

## HI-FIBRE ASTRONOMY

When an experimental fibre-optics system for the UK Schmidt Telescope in north-western New South Wales was proposed more than twenty years ago, it was with velocity surveys of stars in mind. The idea was to use optical fibres to carry light from a few dozen stars in the telescope's 6° field of view, and line them up in a spectrograph, where all the radial velocities could be measured simultaneously. That having been done, the fibres would then be laboriously rearranged by hand to suit the next set of stars, and the telescope moved on to the new field. Even with the tedious reconfiguration of fibres, it offered a speedy method of collecting radial velocities for large numbers of stars.

Fortunately, as soon as this primitive multi-star speedometer appeared in 1985, it was hijacked by the galaxy red-shift brigade – who wanted to perform the same trick on galaxies rather than stars, in order to map out the local Universe. That dastardly deed almost certainly saved it from oblivion. Stellar radial-velocity surveys were very unglamorous in the 1980s, whereas mapping the Universe by measuring the red-shifts of galaxies (and hence their three-dimensional positions in space) was just starting to become the cornerstone of observational cosmology. It was a lucky break for the UK Schmidt, and it was followed by another one in 1988 when the operation of the telescope was taken over by the Anglo-Australian Observatory.

The UK Schmidt is still observing galaxies today, most notably as part of the 6dF Galaxy Survey (6dFGS) being carried out with the telescope's fourth-generation multi-fibre system. That instrument (6dF – for 6° field) allows up to 150 objects to be observed at once, with fibre positioning being carried out by a friendly robot instead of a bored astronomer (Figure 1). By the time the 6dF survey is completed in mid 2005, it will have produced three-dimensional positions for something like 150,000 galaxies over the whole southern sky.

Meanwhile, how many stellar radial velocities do you think the world's astronomers have in their data archives? Astonishingly, it is only about 40,000. Thanks to 6dFGS and other galaxy surveys, we now know far more about the geography of the local and middle-distance Universe than the internal motions of our own Galaxy. And with the Global Virtual Observatory threatening to bring all the world's astronomical archives to a computer near you very soon, this imbalance of data is quickly going to start looking pretty embarrassing.

It is not embarrassment that drives large-scale surveys, however, but scientific impetus. In contrast with the 1980s, there is growing awareness today that many clues to the fundamental problem of galaxy formation in the early Universe are locked up in the motions and chemical compositions of stars in our own Galaxy. It is the idea of tapping into this rich source of information that has spawned ambitious space projects such as the European Space Agency's GAIA. The

**Figure 1.** The 6dF robot positioning optical fibres under the watchful eye of UK Schmidt Telescope observer Kristin Fiegert. The robot positions the fibres so that they will be in alignment with RAVE target stars when loaded into the telescope. (By courtesy of Jonathan Pogson, © Anglo-Australian Observatory.)

prime mission of this proposed satellite is space astrometry – the accurate measurement of star positions unencumbered by Earth's atmosphere. But it will also have a spectroscopic capability to add radial velocities and chemical signatures to its range of measurements. The goals of the GAIA team are not modest – they aim to make these measurements for no less than a billion stars.

Although GAIA's launch is at least seven years down the track – and the first results some years after that – we already have proper motions for several million bright stars in the Hipparcos and Tycho-2 catalogues (118,000 and 2.5 million objects respectively). These observations were made with GAIA's forerunner, Hipparcos, in the late 1980s. Unfortunately, plans to extend them to around 40 million stars with fast-track space missions ahead of GAIA have had to be shelved owing to lack of funding.

Nevertheless, if the existing space-astrometry archives are going to be fully exploited, it is clear that there is a pressing need for radial velocities on an all-sky scale. And if chemical signatures of the target stars can be obtained with the spectrograph at the same time, then so much the better. The most important of these is the so-called metallicity: the ratio of heavy elements to hydrogen in the star's atmosphere, which is an indicator of its age.

In work such as this, field of view is more important than mirror diameter, because the stars are well spread out in the sky but are relatively bright. So it turns out that a large Schmidt telescope is the perfect tool for carrying it out. Especially one with a highly effective multiple-object spectroscopy system that will complete its present task in mid 2005.

## ENTER THE RAVErs

It is this logic that has led to the idea of RAVE. Nice acronym, don't you think? It conjures up all sorts of feverish activity, and certainly matches the fervour with which the project is being conducted. In fact, RAVE stands for RAdial Velocity Experiment, and it is a survey of the radial velocities and metallicities of a large number of stars over the whole sky. It is being carried out with the UK Schmidt Telescope, and was started in April 2003 using nights around each full moon when the telescope is not scheduled for other work – the so-called BOM, or bright-of-moon period. From mid 2005, if all goes to plan, RAVE will use all observing time on the UK Schmidt, with completion of the survey expected around 2012.

Capable though the telescope is, it cannot observe beneath the horizon, and so a second instrument in the Northern Hemisphere will be needed to complete the survey. Current favourite for this is the Calar Alto Schmidt, a telescope with striking similarities to the UK Schmidt but on a smaller scale. The 80-centimetre aperture of the Calar Alto instrument delivers only half the light-grasp of the UK Schmidt's 1.2 metre, and so progress will be slower through the northern portion of the survey. Fortunately, about two-thirds of the sky can be observed from the UK Schmidt's location at Siding Spring Observatory, so the division of work need not be equal.

RAVE is a project of megalomaniac proportions. The pilot survey currently under way will obtain spectra of about 80,000 Southern Hemisphere stars by mid 2005, and is then expected to be extended to around 2007. By that date, well over a third of a million objects will have been observed. But RAVE's ultimate goal is far more ambitious –

namely, to observe the brightest 50 million stars in the Galaxy, with completeness to an *I* magnitude (measured in the far-red waveband) of 15. Little wonder the project team members are sometimes accused of being RAVEing mad!

Those team members are drawn from eleven nations, including Australia and the UK. The Principal Investigator and chair of the RAVE Science Working Group is Professor Matthias Steinmetz of the Astrophysikalisches Institut Potsdam, in Germany. (More details of team members, together with other information about RAVE, can be found on the project's main website at http://www.aip.de/RAVE/.) Large-scale surveys such as this do not happen without substantial resources being devoted to them, and it is expected that the science funding agencies of most of the participating countries will help to pay for it. The Anglo-Australian Observatory and the Australian Research Council have both contributed to get the project off the ground during the period 2003–5.

RAVE uses observations made in the far-red region of the spectrum. They are centred on a trio of dark lines that originate in the element calcium in the stars' atmospheres. The lines of this so-called calcium triplet have accurately known wavelengths (of 849.8, 854.2 and 866.2 nanometres, for the technically minded). As well as being suitable for radial-velocity measurements, the far-red region is well endowed with other diagnostic features suitable for metallicity determination. It is also relatively free from absorption originating in the Earth's atmosphere (so-called telluric absorption), and is less affected by moonlight than the blue part of the spectrum. Observations so far indicate that velocity errors of less than 2 kms$^{-1}$ can be obtained from such data, a precision achieved largely because of the stability of 6dF's spectrograph. This is no accident – the spectrograph sits in its own private room in the dome, with an eleven-metre-long fibre cable connecting it to the telescope's focal surface.

With such high-precision radial velocities, and equivalent accuracy in metallicity and other chemical signatures, it has quickly become clear that RAVE has the potential to be an invaluable stand-alone resource, whether or not it is complemented by space-based astrometric data. In fact, the finished RAVE database looks set to revolutionize our understanding of the formation and evolution of all the major components of the Galaxy: the disk, the bulge and the halo.

What are the scientific questions that RAVE scientists want to address? Many of them centre on the dynamical history of our Galaxy – in other words, the way that large-scale motions of stars have evolved to become the way we see them today. Imagine for a moment our Galaxy as it was, say, 10 billion years ago. We expect from simulations of galaxy formation that its structure would have been very different from what we see now. No elegantly curving arms representing 'grand-design' spiral structure, but instead, an irregularly shaped proto-galaxy surrounded by a rag-tag collection of smaller satellite galaxies. We have some faith in this model of galaxy formation, as there are hints of such untidy structures in the 'deep-field' images provided by Hubble and other telescopes, in which we look back in time perhaps 10 billion years.

What we believe has happened to our Galaxy in the aeons since then is that the satellite galaxies have been gravitationally gobbled up to form various components of today's Milky Way (Figures 2a and 2b). Most notably, we believe many of them disintegrated to form the halo of our Galaxy – that spheroidal mass of stars and globular clusters within which lurks the Galaxy's store of mysterious dark matter. But

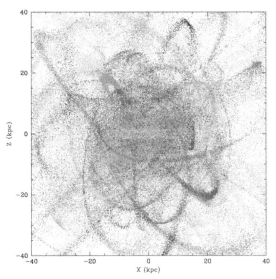

**Figure 2a.** How RAVE will shed light on our Galaxy's history. This simulation shows a galactic halo built up through the accretion of 100 satellite galaxies. The disrupted remnants of the individual satellites are clearly seen. (By courtesy of Paul Harding and Heather Morrison.)

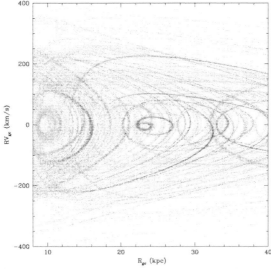

**Figure 2b.** The same simulation plotted in phase-space, i.e. the radial velocity of each star in the simulation is plotted against the radius of its orbit around the galactic centre. RAVE has the potential to be able to distinguish the different orbits very clearly. (By courtesy of Paul Harding and Heather Morrison.)

crucially, the stars that came from these devoured satellites still carry a memory of their origin with them – in their orbits through space. Thus, analysis of the velocities of very large numbers of stars allows a kind of galactic archaeology to be carried out. Stars with a common origin can be disentangled from the mass by their common velocities – and also their similar chemical signatures.

In the broadest sense, RAVE is designed to allow a comparison of what we see in our Galaxy today with what simulations of galaxy formation predict. Theoreticians have provided sophisticated models involving millions of test particles – artificial stars – and naturally want to compare them with reality. The problem is that at this level of detail, we don't know what reality looks like – hence the need for RAVE.

Other questions the survey will address include the chemical history of the Galaxy, the origin of the so-called 'thick disc' (a rarefied layer of stars above and below the main disc), and the origin of the bulge at the Galaxy's nucleus – that blaze of stars glimpsed through intervening dust clouds on Southern Hemisphere winter nights. The idea of 'weighing a spiral arm' (by analysing the motions of stars within it) is also one that has caught the imagination of RAVErs, and no doubt it will be one of the first significant results to come from the survey.

## RAVE'S PROGRESS

At present, the RAVE pilot survey is being conducted with 6dF on the UK Schmidt Telescope during seven nights per lunar month. By mid 2005, when the 6dF Galaxy Survey is completed, it is hoped that will increase to twenty-five nights per month, with RAVE taking up all the time on the telescope. But there is clearly a problem here. Using 6dF, about 700 stars can be observed on an average clear night, and it doesn't take a genius to work out that at this rate, the UK Schmidt's share of RAVE (about 30 million stars) will take something like 210 years to observe – with due allowance for cloudy weather. That is rather longer than most RAVErs expect to have available.

The solution to this problem is waiting in the wings. It is a new instrument that will one day replace 6dF on the UK Schmidt, an instrument whose capabilities will make 6dF seem as crude as its manually operated predecessor does today. This is *Ukidna*, the holy grail of wide-field, multi-fibre spectroscopy.

The name is derived from the echidna, Australia's marsupial spiny anteater, and this hedgehog-like creature provides rather a good illustration of the working of the instrument. Imagine something like an echidna held upside-down in the focus of the Schmidt Telescope's concave mirror. (Forget animal cruelty for a minute – this is just a thought experiment.) That position is where an image of the sky is formed, and if the spines of the echidna each had an optical fibre running down the centre and could be independently tilted to a predetermined position, then the obliging echidna would allow you to align many fibres with target objects on the sky.

Such an instrument (not surprisingly christened *Echidna*) is currently being built by the Anglo-Australian Observatory for the Japanese 8-metre Subaru Telescope at Mauna Kea in Hawaii (Figure 3). Its robotic technology is reasonably well established, and is expected to be adaptable to the UK Schmidt. The main difference is one of scale: whereas Echidna on Subaru has 400 fibre-carrying spines, Ukidna on

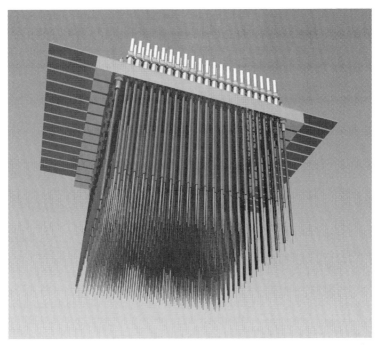

**Figure 3.** Simulated view of the Echidna fibre positioner on the Subaru telescope. The fibre spines are individually adjusted by robotic control so that the lower end of each one intercepts the light from a target star or galaxy. (By courtesy of Andrew McGrath, © Anglo-Australian Observatory.)

the Schmidt Telescope will have no less than 2,250 – all of which have to be stirred into robotic action at the same time. And each fibre has to feed back from its spine to a spectrograph mounted in that private room we mentioned earlier, where all 2,250 spectra will be recorded simultaneously. This is challenging technology, but there is no reason to believe it cannot be brought to reality. And when it eventually is, the UK Schmidt Telescope will be able to record the spectra of some 22,000 stars per average clear night, allowing the main part of the RAVE survey to be completed in something like six years.

The bigger challenge at present – certainly for yours truly – is trying to work out how Ukdina can be funded in time to be operational by 2007. Not only will the complex mechanical hedgehog have to be built, but also a new spectrograph to record the results. And then the cost of those needs to be doubled, because the same thing would have to be built for the Calar Alto Schmidt, if that does turn out to be RAVE's Northern Hemisphere instrument. Add to that total the running costs of the two telescopes, and you are soon talking many millions of dollars, pounds, euros or yen. At present, most RAVErs are optimistic about raising the appropriate funding, such is the scientific urgency of the project. But only time will tell.

By the middle of 2005, the RAVE pilot survey in the capable hands of the UK Schmidt Telescope's observing team will have tripled the world's store of stellar radial velocities. That work is already funded, and even if the survey were to proceed no further, it would be a remarkable achievement for one of the world's most unassuming telescopes. But if, as everyone hopes, the full RAVE survey proceeds beyond 2007, it will provide a spectacular finale to the UK Schmidt's long career. And the omens are good. A couple of years ago, when two Anglo-Australian Observatory astronomers were first hatching the idea of Ukidna during the long drive from Sydney to the telescopes, they rounded a bend in the road to come face to face with – an echidna. Screeching to a halt, they gently moved the animal out of harm's way, whereupon it scuttled off into the undergrowth. If you believe in good signs, they don't come any better than that!

And so to one final thing – that gobbledygook sentence we started with. Just in case you didn't manage to decipher it, the corrected version is:

Smith, where Jones had had 'had', had had 'had had'; 'had had' had had the approval of the examiners.

Like stars without velocities, words without punctuation are just about enough to drive you RAVEing mad!

## FURTHER READING

Freeman, K. and Bland-Hawthorn, J. (2002), *Annual Review of Astronomy and Astrophysics* 40, 487.

Steinmetz, Matthias (2003), in Ulisse Munari (ed.), *GAIA Spectroscopy, Science and Technology*, ASP Conference Series, 381–6.

Steinmetz, M. and Navarro, J.F. (2002), *New Astronomy* 4/7, 155.

See also the following internet sites:

RAVE: http://www.aip.de/RAVE/
6dF: http://www.aao.gov.au/ukst/6df.html
Ukidna: http://www.aao.gov.au/local/www/schmechidna/

# The Mars Revealed by Mariner 4

## DAVID M. HARLAND

### EARLY IDEAS ABOUT MARS

When Mars was at perihelic opposition in 1877, Giovanni Virginio Schiaparelli in Milan mapped the ochre and dark patches on its surface. The most astonishing features on his map, however, were a network of bands crossing the ochre areas that he referred to as *canali*, meaning channels, and which appeared to have been 'laid down by rule and compass'. This announcement prompted a wave of interest in the planet, and over the next few years people reported all manner of unusual phenomena. Percival Lowell, an American millionaire, built an observatory to study the planet at the favourable opposition of 1894. In a book entitled *Mars* published the following year, he drew a map bearing an elaborate network of fine lines, and concluded:

> Firstly, that the broad physical conditions of the planet are not antagonistic to some form of life; secondly, that there is an apparent dearth of water on the planet's surface and, therefore, if beings of sufficient intelligence inhabit it, they would have to resort to irrigation to support life; thirdly, that there turns out to be a network of markings covering the disk precisely counterparting what a system of irrigation would look like; fourthly, and lastly, there is a set of spots placed where we should expect to find the land thus artificially fertilized, and behaving as such constructed oases should.

The prospect of a dying Martian civilization prompted H.G. Wells to write *The War of the Worlds* in 1898. Although astronomers rejected Lowell's thesis, the public was captivated. However, as estimates of the thickness of the planet's atmosphere fell, the prospects for life diminished. By the early 1960s, the atmosphere was believed to have a composition similar to our atmosphere, and a surface pressure of 100 millibars, but the measurements were difficult. A telescope flown on a stratospheric balloon in 1963 verified the presence of carbon dioxide and water vapour at a pressure of about 10 millibars, but failed to detect the nitrogen that was believed to form the bulk of the atmosphere. Even although there were no intelligent Martians, it was nevertheless widely believed that the seasonal variations in the dark areas were due to a primitive form of life akin to lichen.

The advent of the space age presented the opportunity of sending instrumented probes to make close-up observations of the planets, to discover what the telescope could not reveal. The Soviets were first to try to send probes towards Mars, but their early efforts were foiled by technical problems, and the first probe to make even a start on the long journey (launched in November 1962) fell silent en route. Having succeeded in flying a probe by Venus in 1962, the Americans decided to try their luck with Mars.

## A PROBE TO MARS

Built by the Jet Propulsion Laboratory, a part of the California Institute of Technology in Pasadena, near Los Angeles, and operated by NASA, Mariner 4 was constructed around an octagonal framework (Figure 1). Because it was three-axis stabilized in space, the elliptically shaped dish of the S-Band high-gain antenna (which was 1.15 metres across its major axis and 0.5 metres across its minor axis) was mounted on the top of the frame in a position such that the Earth would be within its narrow beam during the probe's encounter with Mars. From the centre of the dish projected a 2.2-metre-long tube, on the tip of which was the low-gain antenna that would receive commands from Earth. Seven of the enclosed bays of the frame held spacecraft and scientific systems; the eighth held the liquid-propellant rocket motor for the mid-course correction, whose nozzle was on the side panel. A cruciform of 0.9-metre by 1.8-metre solar panels carried a total of 28,224 transducers to generate some 300 watts of power at Mars's distance from the Sun. On the tip of each panel was a small vane that the attitude control system could tilt to correct any imbalanced forces imparted on the probe by the plasma of the 'solar wind'. In its deployed configuration, the probe spanned 6.8 metres and was 3.3 metres tall. Of its 260-kilogramme mass, some 27 kilogrammes was scientific instruments and associated systems. It had six particles and fields experiments to study the solar

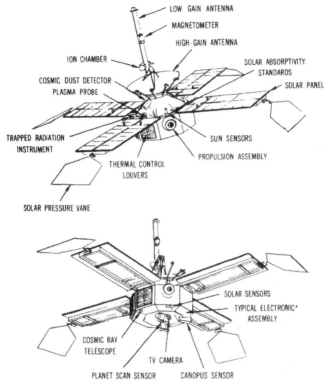

**Figure 1.** A diagram of the main structural elements and instruments of Mariner 4.

wind en route to Mars and in the planet's local environment, and a system to image the planet during the all-too-brief fly-by.

Launch windows for Mars occur every twenty-five months. Two identical spacecraft were built for the 1964 window. Mariner 3 set off on 4 November. The Atlas booster worked perfectly, but the aerodynamic shroud of the Agena-D 'escape' stage failed to separate, trapping the probe. An investigation discovered that the new lightweight shroud had bonded to the probe. A rapid revision of the shroud enabled Mariner 4 to be launched on 28 November, a few days before the window expired. As soon as the probe was safely on its way, it hinged its solar panels down into the plane of the octagonal frame, and oriented itself to face the panels towards the Sun to recharge its battery. Early in the interplanetary cruise the star-tracker employed for attitude determination kept losing its lock on the bright southern star Canopus, and the solar plasma detector malfunctioned, but these problems were overcome.

There were several rules for determining the spacecraft's trajectory during the encounter (Figure 2). Firstly, neither Mars nor its two moonlets (Phobos and Deimos) could be permitted to block the star-tracker's view of Canopus. Secondly, the probe could not fly through the shadows of any of these bodies. The trajectory was designed to provide a slow fly-by, 240 days into the mission. As the probe caught up with the planet in its orbit, its observations would be of the trailing hemisphere. The post-encounter trajectory was required to take the probe behind the planet (as viewed from Earth) so that the manner in which the radio signal was cut off by the limb would give the first definitive measurement of the density of the planet's atmosphere. A second measurement would be made when the probe emerged from the leading limb. Crucially, this occultation had to be timed for when Mars was above the horizon of the radio receiver at the Goldstone tracking station in California. A mid-course correction on 5 December trimmed the initial 240,000-kilometre miss distance down to the intended 10,000-kilometre fly-by, set up this timing, and thus determined the features that would be in the camera's field of view. It had initially been hoped to get a look at the prominent dark feature Syrtis Major, but this was ruled out by the delayed launch. In general,

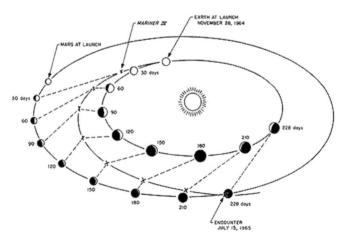

**Figure 2.** Mariner 4's trajectory from Earth to Mars.

the objective of the mission was to secure pictures comparable in resolution to contemporary telescopic pictures of the Moon.

## IMAGING THE RED PLANET

On the base of the octagonal frame was a vidicon imaging system on a platform that could be rotated through 180°. The system was powered up six hours before the encounter in order to give JPL time to trouble-shoot any problems (diagnosing a fault and issuing recovery actions would have to be prompt, because the light-speed travel time over such a distance was twelve minutes). With one hour to go, the platform was rotated until its wide-angle sensor noted the planet's presence, at which time it centred the camera on the planet. The imaging sequence was started when the narrow-angle sensor detected the planet, about half an hour later. (As a back-up measure, the imaging system, which was 'looking', would have self-triggered as soon as the planet's disk encroached on its field of view.) A television camera displayed the image from a 4-centimetre-diameter aperture f/8 Cassegrain telescope with an effective focal length of 30 centimetres on to a small phosphor screen, and this was digitized into an array of $200 \times 200$ pixels, each of which encoded a 6-bit greyscale measurement. The data from the particles and fields instruments were transmitted in real time, but the images were stored on a 100-metre loop of magnetic tape. The exposure time had been pre-set at 1/20th of a second, this being the best bet based on expected illumination. It took 24 seconds to read out the image from the vidicon and another 24 seconds to clear the screen, so images could be taken no more rapidly than once every 48 seconds. The system had a rotating shutter with alternating blue-green and orange-red filters, these having been selected to enhance the contrast in the greyscale and to emphasize the colour differences of the albedo features on the Red Planet.

Several minutes after the point of closest approach, the probe crossed the planet's orbit, and an hour and a half later it was occulted. This revealed that the atmosphere was composed primarily of carbon dioxide with some water vapour and a trace of argon, with a pressure at the surface of 10 millibars. It was belatedly realized that the best terrestrial measurements had been accurate, and that what had been detected *was* the atmosphere – it was not 'bulked up' by nitrogen.

There could be no liquid water on the surface at such a low pressure, and this in turn suggested that the polar caps were mainly frozen carbon dioxide rather than water ice (an assumption only recently shown to be false). The other instruments established that if Mars has a magnetic field it is exceedingly weak, and unable to form a magnetosphere to ward off the plasma of the solar wind, which must interact with the atmosphere. The fact that the surface would be irradiated by solar ultraviolet and cosmic rays made the likelihood of a form of life even as 'simple' as lichen most unlikely.

About 12 hours after the occultation, the tape began to replay the images. At 8.33 bits per second using the 10-watt transmitter, it took 8 hours and 20 minutes to transmit the 240,000 bits in each image. As each frame was followed by engineering data, the effective rate was one frame every 10 hours. It took 10 days to replay the tape. It was promptly replayed to enable any transmission flaws to be identified and eliminated.

Taken from a range of 16,500 kilometres, the first frame showed the planet's limb and the black of space. The engineers were ecstatic, since this confirmed that the camera had properly acquired the planet. As the other frames were processed, however, there was disappointment. Despite having employed filters to highlight the contrast on the planet's surface, most of the twenty-two images were bland due to 'flare' in the optics, and the rest were black. However, JPL had a computer program to 'enhance' digital imagery by adjusting the shades of grey to 'stretch' the contrast range, and once this was done some surface details were revealed.

## A SURPRISING DISCOVERY

The imaging sequence (Figure 3) began with a view looking northwards across the limb at a longitude of about 190° (planetary longitudes are measured westwards from the prime meridian, which in the case of Mars is defined as passing through a small dark feature named, appropriately, Sinus Meridiani), and traced south-east and across the equator to about 52°S, where it curved north again and crossed the terminator prior to departing the planet in darkness at about 38°S, having spanned about 100 degrees of longitude. In the first few frames the surface appeared to be blotchy, with hints of large circular features.

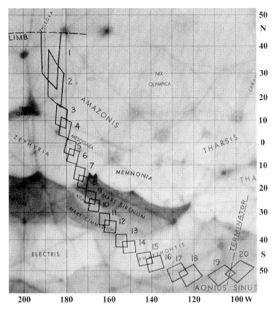

**Figure 3.** The areas covered by the sequence of frames taken by Mariner 4, overlaid on a section of a map of Mars drawn by telescopic observers. Frame no. 1 looked northwards across the limb over a 'desert' feature known as Phlegra at longitude 190°. Frames 7, 8 and 9 spanned the north-western part of the dark albedo feature Mare Sirenum. Frame no. 10 was in a bright strip of Atlantis. Frames 11 and 12 spanned the dark strip of Mare Cimmerium. As the track began to swing east, the next few frames crossed the desert of Phaethontis. The next frames, running east at about 50°S towards the terminator, were featureless. Note that many of the names assigned by telescopic observers have since been deleted.

On frame no. 7 it became apparent that these were impact craters. This came as a surprise. It was widely accepted that the Moon's surface was cratered because it bore the scars of its violent formation, but Mars had been expected to be an 'evolved' planet, similar to the Earth. The presence of craters meant that its surface was ancient – as old, in fact, as the lunar surface. The craters were shown with greater clarity with each successive frame. The crater 120 kilometres in diameter on frame no. 11 (Figure 4), which was taken from a range of 12,500 kilometres, has become the iconic Mariner 4 view (and has since been named 'Mariner' in honour of the probe). The last frame on which surface detail could be seen was no. 15. Although the contrast on the terminator would have emphasized the topography, the sensor that was to have increased the exposure as the field of view darkened had failed. The final few frames were snapped beyond the terminator.

Although the imaging system was marred by flare, it *had* achieved its primary objective of revealing the nature of the Martian surface. Altogether, the imaging track covered barely 1 per cent of the planet, but the presence of craters on a variety of albedo features suggested that the entire surface was ancient. Moreover, the fact that the craters had retained their form for billions of years implied that there had been little or no erosion – Mars, now cold and dry, had apparently never been sufficiently wet for a hydrological cycle in which rainfall chemically and mechanically eroded the surface. Furthermore, the fact that the craters had retained their shape indicated that the surface is not subjected to the tectonism that continually reshapes terrestrial topography.

Later spacecraft may have given us far more detailed views of Mars, and indeed of Mariner Crater (Figure 5), but in giving us our first close look at the Red Planet, Mariner 4 stands out in the history of Solar

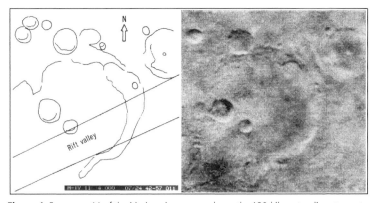

**Figure 4.** Frame no. 11 of the Mariner 4 sequence shows the 120-kilometre-diameter crater that was named in honour of the probe. Space historian Eric Burgess pointed out that this crater lies on the line of one of the *canali*, and noted that it is crossed by a number of parallel linear features that he interpeted as grabens – long thin strips of land that dropped between parallel faults opened when the crust was subjected to extensional stress. If this was the case, then there had been *some* tectonic activity since the ancient period of crater formation.

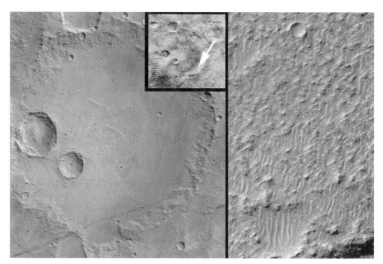

**Figure 5.** Viking Orbiter documented Mariner Crater in greater detail (left), and Mars Global Surveyor snapped the floor of the crater and the edge of one of the linearities (arrowed in the Mariner 4 context view), which is indeed a graben.

System exploration – no robotic probe, before or since, has had such a dramatic impact. The announcement by Percival Lowell that Mars was inhabited by an ancient race engaged in a struggle to survive as their world dried up failed to impress the astronomers, but it inspired epic storytelling by Edgar Rice Burroughs, Ray Bradbury and Arthur C. Clarke, among others. Then, on 14 July 1965, even the prospect of lichen on Mars was dismissed as fantasy. In the light of its tenuous atmosphere and seemingly inert ancient surface unweathered by water, the *New York Times* made a play on Mars's moniker of the 'Red Planet' by referring to it as the 'Dead Planet', and almost overnight scientific interest in Mars plummeted.

Of course, subsequent research has shown that some of the conclusions drawn from the flight of the Mariner 4 were wrong (see Peter J. Cattermole's article on page 154), but it was Mariner 4 that showed the way, and it remains one of the great scientific triumphs of the twentieth century.

## FURTHER READING

Crosswell, Ken (2000), *Magnificent Mars*, Free Press.

Godwin, Robert (ed.) (2000 and second volume 2004), *The NASA Mission Reports*, Apogee Books.

Hanlon, Michael (2004), *The Real Mars*, Constable & Robinson.

Hartmann, William K. (2003), *A Traveler's Guide to Mars – The Mysterious Landscapes of the Red Planet*, Workman.

Morton, Oliver (2003), *Mapping Mars – Science, Imagination and the Birth of a World*, Fourth Estate.

# The Music of the Spheres

## ALLAN CHAPMAN

Most people know that, in the early seventeenth century, the German astronomer Johannes Kepler suggested that the orbits and velocities of the planets moving around the Sun could be expressed in accordance with musical proportions. Yet the idea of 'the Music of the Spheres' was already over 2,000 years old by the time that Kepler wrote his *Harmonici Mundi* ('Harmony of the Spheres') in 1619. Of course, Kepler was familiar with these ideas because, in addition to being one of the greatest mathematical astronomers of all time, he was also a deeply learned classical scholar and linguist who was well acquainted with the philosophical ideas of the ancient Greeks.

Yet why should anyone assume that there was any kind of connection between the sounds and pitches of the musical scale and the heavens? The answer lies in earlier beliefs in the nature of mathematics, the changeless perfection of the heavens, the instinctive preference of the human mind for the harmonious and logical, and the idea that all were bound together in a single divine plan. In a nutshell, the proportions present in both music and in mathematics meant that the human mind had access to a technique whereby it could understand the very structure of the cosmos.

## THE IDEAS OF THE GREEKS

According to Greek legend, it was Pythagoras, who lived in the sixth century BC, who first came up with the idea. It is said that one day, this 'Father of Greek Astronomy and Mathematics' noticed a blacksmith at work. When the blacksmith struck his anvil with a hammer of a given weight, it always produced a single tone; quiet or loud, depending on the force of the blow, but *always* the same tone. Heavier hammers, moreover, always produced deeper notes than light ones. Pythagoras had found that there was an unchanging relationship between hammer weight and sound. And what is more, pipes of a given length, such as those in a set of pan pipes, always gave the same note, no matter how hard or soft one blew into them. And why did the Moon always seem to go around the Earth in 28 days, Mercury in just over 80, the Sun in 365.25 and Jupiter in 12 years?

Yet there was another powerful consequence that fascinated the ancients. Just as the open musical scale contained seven notes through which sequence the pitch returned to the octave, so there were seven 'planets': the Moon, Mercury, Venus, the Sun, Mars, Jupiter and Saturn, with the stars forming the celestial octave.

Now in classical Greek times, it was believed, from the best available observational evidence, that the Earth was fixed in the geometrical centre of the Universe. And around the spherical Earth was a series of 'crystalline' spheres – popularly said to be nine in number – that nestled inside each other like the skins of an onion. On each of the seven innermost spheres were the seven 'planets', the Moon to Saturn, one planet to each sphere. What is more, these seven planetary spheres were perfectly transparent, thus enabling people on Earth to see through and beyond them. The eighth sphere, however, was black and opaque, although it carried the fixed stars and the Milky Way, and explained why the night sky was dark. Then beyond these visible spheres was an invisible ninth sphere that acted somewhat like a great flywheel to all the rest.

And just as a hammer on an anvil, a blown pipe, or a plucked string always emits a single tone, or pitch, that corresponds to its weight or length, so each planetary sphere always rotates around the Earth (as they believed) at a given speed: thus, the Moon 28 days, Mercury 80 days, Venus 270 days, Sun 365.25 days, Mars 2 years, Jupiter 12 years, Saturn 29.5 years; and the 'octave' fixed stars, once every terrestrial day. This perceived congruence between the physics of sound and the motions of the heavens was one of the first triumphs of scientific thinking, for no one before the Greeks had seen nature in such an interconnected way.

This Greek fascination with mathematics, astronomy, music and symmetry, moreover, was not just confined to external objects, such as short and long pipes and planetary periods; it was also seen as applying to the very intelligence of the human observer. Just as the strings and pipes of the earthly musical scale formed a harmonic sequence, so the motions of the planetary spheres did likewise. And here, the Greek philosophers such as Pythagoras and Plato saw a connection to what we would now call psychology. Why do we find harmonious sounds enchantingly beautiful, and why do elegant geometrical forms give us such delight? And why, conversely, do dissonances make us grind our teeth, and lopsided, asymmetrical shapes and structures strike us as ugly? Plato said that it is because our minds are designed to respond to eternal truths and beauties, and that these beauties are innately geometrical, harmonious and celestial. Hence, to the Greek philosophers, number, harmony, cosmology and intellectual and sensory delight are all intimately connected, and mankind feels driven to study the heavens because we all aspire to understand and enjoy these harmonies. And by the same logic, we are repelled by the unharmonious and the irregular.

These pagan Greek ideas about number and harmony passed into medieval Christian, Jewish and Islamic culture, and entered the curricula of Europe's great medieval universities. The 'music of the spheres' was lectured upon in twelfth-century Paris, Bologna, Oxford and elsewhere as part of the 'Quadrivium' course, where Arithmetic, Geometry, Music and Astronomy passed on the Greek culture of proportion to every undergraduate across a span of more than six centuries. Indeed, the widespread assumption that astronomical and mathematical subjects were not taught in the universities of medieval Europe is a folk myth that deserves to follow the Bogeyman and the Walking Dead into the graveyard of historical untruths.

## THE INSIGHT OF KEPLER

By the early seventeenth century, therefore, the music of the spheres was a firmly established concept among educated Europeans. Nor should we forget that it was also being developed in England by philosophers and writers such as Robert Fludd (Figure 1). But how did Kepler in particular succeed in giving it such a new lease of life?

By 1600, astronomy was developing faster than at any time in the previous 2,000 years. Based upon his analyses of the unprecedentedly accurate observations of Tycho Brahe, and in particular of Tycho's observations of the motions of the planet Mars, Kepler came to the conclusion that the idea published by Nicholas Copernicus in 1543 that the Earth and the planets rotated around the Sun (and not the Earth) was indeed correct.

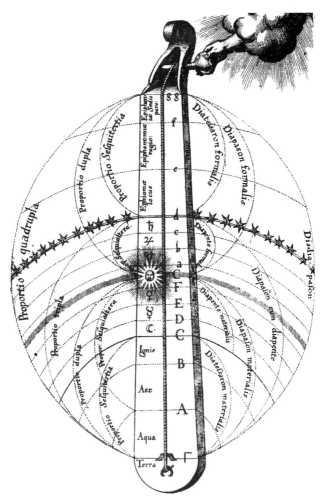

**Figure 1.** In the early seventeenth century, the music of the spheres was developed in England by writers and philosophers such as Robert Fludd (1574–1637), whose *Divine Monochord* (1617) demonstrated a string being tuned by the hand of God to produce the musical harmonies of the cosmos.

Yet if the Earth and the planets moved around the Sun, and comets seemed (as Kepler's contemporaries were coming to believe) to have some orbital relationship with the Sun instead of being generated in the Earth's atmosphere, then why did the crystalline spheres not get in the way? Indeed, Johannes Kepler was one of the first astronomers to formally abandon the ancient crystalline spheres, suggesting instead that the spinning Sun gave off some kind of magnetic flux that carried all the planets around in one orbital plane and in one direction. In so many ways, however, it was the wonderful realization of these discoveries – Copernicanism, an invisible solar force that carried the planets through empty space, and perhaps most of all his realization that at least Mars moved around the Sun in an *ellipse,* and not a *circle* - that gave the music of the spheres a new and potent meaning to Kepler. For Kepler was the antithesis of a materialist scientist. He was a mystic, and a deeply religious man who believed that the cosmos was bound together by a whole hierarchy of spiritual and philosophical truths that constituted the greater reality. And to Kepler, mathematics, harmony and a celestial musical geometry lay at the heart not only of planetary motion but also of Truth itself.

And why, moreover, was there a *trinity* of laws of motion, one, two, three? Why did the force that acted between the Sun and a planet not merely weaken with distance, but seemed to do so in accordance with a precise mathematical rule? And why did this force make planets move not in circles or parabolae, but only in ellipses: speeding up as they approached the Sun, and slowing down as they receded? And how, in this new, empty universe, with no solid spheres, but only invisible forces binding everything together, could one calculate the endless speedings up and slowings down? Quite simply, one needed a new mathematical language, and it was in accordance with this need that Kepler began to seek fresh inspiration in the music of the spheres.

By the time he began to wrestle with the mathematical problems of elliptical orbits, around 1603–9, Kepler had long been familiar with the geometrical and physical problems of the Copernican universe. Indeed, one day, over a decade earlier, while delivering a mathematical lecture at the University of Graz, on 9 July 1595, it is said, it suddenly hit him in a flash that one could go some way to describing the observed eccentricities of the planetary orbits in Copernican astronomy if one imagined each orbit to be moving within the spatial volumes of particular regular solids, such as those described by the Greek geometers:

solids such as the cube, tetrahedron, dodecahedron, icosahedron and octahedron. This realization, in fact, had already won him his first European-wide fame when, at the age of 25, Kepler published these ideas in his *Mysterium Cosmographicum* (1596). It was also this work that led to Kepler being noticed by Tycho Brahe.

When Kepler was making his first discoveries in planetary geometry, the techniques of mathematical analysis were still at a fairly basic level. Algebra, which had been invented in the Arab world, was still scarcely understood, while logarithms, decimal notation and equations were still in their infancy. And one part of mathematics that was still something of an intellectual wilderness was the fractional or 'infinitesimal' analysis of curves. In short, how could one break up an elliptical orbit into a series of fractional units, and correctly express the changing velocity of a planet's motion as it passed within that orbit? And how could those fractional units be related to the Sun, which occupied only *one* of the ellipses's two foci (the other being empty)? This problem was to plague mathematicians for a good bit of the ensuing seventeenth century, though it fell to Kepler's genius to realize by about 1619 that one might use *musical* language to express them.

In his *Harmonici Mundi* (1619), Kepler developed the idea of expressing the properties of a planet's orbit in terms of musical notation (Figure 2). Mercury, which in the Copernican system was the planet closest to the Sun, was expressed in rapidly moving notes, up and down the musical stave an octave above middle C. Venus, the orbit of which is almost circular, is expressed in one high tone, whereas Mars, the complex orbit of which had baffled Tycho Brahe and the young Kepler, was represented by a tonal frequency that moved up and down across five notes. And Saturn, the dull, slow-moving planet that was thought in 1610 to be at the very edge of the Solar System, has its orbital characteristics represented by three tones, deep down in the bass part of the scale: slow-moving, leaden and ominous, like the low pedal notes on the organ. Of course, Kepler did not believe that the planets actually emitted these musical notes in an audible sense; it was just that he used the mathematical flexibility of musical notation to express these very complex and ever-changing curvilinear proportions that a planet describes as it rotates around the Sun in an ellipse.

Yet we must not forget that astronomy to Kepler was not just physical; it was also intimately bound up with his deeply mystical religious faith. For in his world view the planets did in fact possess a

**Figure 2.** In his *Harmonici Mundi* (1619), the German astronomer and mathematician Johannes Kepler developed the idea of expressing the properties of a planet's orbit in terms of musical notation.

mysterious musicality, as all their orbits blended together in one glorious coherence – a true 'Harmony of the Cosmos', no less – while at the same time this harmony struck a chord of wonder in the rational intellect of mankind.

Johannes Kepler's insights into planetary dynamics were to inspire many other seventeenth-century astronomers, including Jeremiah Horrocks, John Wallis, Pierre Gassendi, Christiaan Huygens, Robert Hooke, Gottfried Leibniz and Sir Isaac Newton. In their hands, mathematical astronomy was to progress immensely between 1620 and 1700, transforming geometry and algebra, and developing new analytical techniques. And by the time that Newton published his *Principia Mathematica* in 1687, it had been demonstrated that the force that held the planets in their orbits was not some sort of magnetism so much as universal gravitation. We must not, however, forget that Newton himself was no less of a mystic than Kepler had been, and was equally enchanted and inspired by the heavenly harmonies of proportion and perfection that he saw as lying at the heart of astronomy.

## SIR WILLIAM HERSCHEL: MUSICIAN AND ASTRONOMER

Whereas the mathematical properties of music might have inspired philosophically minded astronomers, we must not forget that over the last 400 years many musicians have also been inspired by astronomy. And in the same way as mathematical astronomy advanced rapidly in and after Kepler's time, so music itself passed through major developments as a performance art. Modern systems of tuning and pitch came into use during Kepler's lifetime and just after, along with opera, oratorio, complex instrumental ensemble music, large organs and sophisticated musical compositions. Indeed, Johann Sebastian Bach and George F. Handel were both born in the year 1685, at the same time as Newton was working on his *Principia*. And all of them, in the early eighteenth century, would be fascinated in turn by the idea of the celestial harmonies, and would try to capture them in their glorious musical compositions. And of all the musical forms, the art of fugue writing was the most harmoniously and mathematically complex.

But in Sir William Frederick Herschel (as he anglicized his name from the German) we find an example of a successful and accomplished art-musician, whose music acted as a vehicle whereby he was to commence one of the most remarkable astronomical careers of all time.

Sir William Herschel was born into a musical family in Hanover in 1738, and after service as an army bandsman came to seek his fortune in England. He thrived there, and brought over his sister Caroline, while his brothers Jacob and Dietrich also performed in England. For England in the 1770s was the most prosperous, most peaceful and most individually free country in the world. It was also a land of opportunity for gifted European musicians, for the theatres, opera houses, concert halls, pleasure gardens and churches of London, Bath, York and the other great cities of England generated an endless demand for musical talent. William Herschel grew rich and successful in Bath. In his leisure time, he began to study the physics of music and sound, as they were then understood, yet after studying Dr Robert Smith's *Treatise on Harmonics* (1749) he went on to read Smith's *Treatise on Opticks* (1738), and this triggered his serious interest in astronomy. He quickly discovered an innate genius for the polishing of mirrors for reflecting telescopes, and after he discovered the planet Uranus in March 1781, Herschel found himself propelled into international scientific fame. King George III invited him to give up music and become the King's Astronomer *(not,* as is sometimes erroneously stated, Astronomer Royal), and from 1782 onwards he embarked upon a career of cosmological discovery that would extend down to his death in 1822.

Sir William Herschel's greatest contributions to astronomy lie in his discovery and classification of several thousand nebulae and double stars. Yet how far away could these dim objects be, Herschel asked? Working on the assumption that a ray of light became proportionately dimmer the further it travelled, he came to the conclusion that these nebulae were vastly more remote than the stars that we see in the ordinary constellations.

But what Herschel was doing was applying a musician's understanding of proportion and mathematical ratios to 'gage' or gauge the universe, as he put it. For in the same way that, as the ancient Greeks knew, vibrating strings and pipes produced mathematically precise pitches, depending on their lengths, so might it be possible to establish the proportionate distances of dim nebulae by 'tuning' or estimating the strength of their light with relation to the brighter stars.

We now know that Herschel's hopes for understanding the structure of deep space were rather optimistic, considering the techniques of 200 years ago, although he was right in principle. And when modern astronomers use the spectroscope, photographic imaging, radio and other techniques to gauge the universe they are using proportionate techniques that relate to musical ways of thinking.


## RECENT DEVELOPMENTS

Nineteenth- and twentieth-century Europe and America witnessed unprecedented developments not only in astronomy, but also across the whole range of the sciences. Similarly, the art of music expanded as never before, becoming increasingly related to advances in technology from the actual improvement to the sounds produced by the musical instruments themselves, to the development of recording and electronic broadcasting techniques to carry music to millions of people worldwide.

And just as the awe-inspiring vastness of Herschel's cosmology captured the imaginations of creative artists during the Romantic era in the early nineteenth century, so the relativistic cosmos of Albert Einstein (himself an amateur violinist) and the galactic discoveries of Edwin Hubble did likewise in the twentieth, not to mention re-awakening mankind's ancient association of the planets with particular human attributes, such as intelligence, love, war, jollity and old age in the creative imagination. But no twentieth-century astronomically related musical composition compares in popularity with Gustav Holst's *The Planets* suite of 1916. For in addition to the ancient planetary associations – especially the ominous staccato of Mars at the time of the Second World War – Holst later became interested in the 'New Cosmology', and read Sir James Jeans's popular works on Einstein, relativity physics and Hubble's 'Island Universes' cosmology: a cosmology, indeed, that would lead to the idea of the Big Bang theory in the 1940s.

In more recent times, moreover, some classical musical compositions came to take on originally unintended astronomical connotations, as when in 1968 Stanley Kubrick's film *2001, A Space Odyssey* used as its theme music Richard Strauss's late nineteenth-century tone poem *Also Sprach Zarathustra*. Nor should we forget that Sir Patrick Moore himself has been active in music as well as in astronomy, producing his opera *Perseus,* and other compositions with astronomical themes.

But lying at the heart of astronomy's relation to music are the Greek concepts of reason, harmony and balance. When astronomy began to advance by leaps and bounds in the seventeenth century, researchers in other sciences began to wonder whether similar divine proportions might also underlie chemistry, medicine and the other sciences. And so they were found to do. For do not John Dalton's ideas of fixed-ratio atomic bonding (1803) and Mendeleev's Periodic Table (1859) in chemistry show that the building blocks of matter combine in ratios that have the elegance of harmonies? And when we think of the DNA code, which governs the mechanisms of cell replication and the physical development of living things, do we not find heavenly harmonies in its vast, multi-layered and elegant complexity?

One has a great deal to thank music for when it comes to the inner logic of the sciences. For when Thales, Pythagoras and their fellow

Greeks in the sixth century BC noted the similarities that existed between the elegances of music and astronomy, they cannot have imagined what their ideas would set in motion. One might suggest that, without the insights of music, neither astronomy nor any of the other sciences would have developed in the way they did. For science, after all, is about seeking out the grand harmonies within nature.

## ACKNOWLEDGEMENT

I should like to thank Donald Francke, performing musician and astronomer, for his insights into the music of the spheres that I gained during our conversation.

## FURTHER READING

Caspar, Max (1993), *Kepler*, 2nd Eng. edn, Dover, New York.

Godwin, Joscelyn (1979), *Robert Fludd, Hermetic Philosopher and Surveyor of the Worlds*, Thames and Hudson.

Gouk, Penelope (1989), 'The Harmonic roots of Newtonian science', in J. Fauvel, R. Flood, M. Shortland, and R. Wilson (eds), *Let Newton Be!*, Oxford University Press, Oxford, pp. 101–26.

Gouk, Penelope (1999), *Music, Science, and Natural Magic in Seventeenth-Century England*, Yale University Press, New Haven and London.

James, Jamie (1995), *The Music of the Spheres. Music, Science, and the Natural Order of the Universe*, Copernicus, Springer Verlag, New York.

Plato, *The Republic*, F.M. Cornford (trans. and ed., 1941, 1966), Oxford University Press, Oxford.

Plato, *The Timaeus*, Desmond Lee (trans. and ed., 1971), Penguin, Harmondsworth.

Stephenson, Bruce (1994), *The Music of the Heavens: Kepler's Harmonic Astronomy*, Princeton University Press, New Jersey.

Tillyard, E.M.W. (1943), *The Elizabethan World Picture*, Penguin, Harmondsworth, 1968.

Van Helden, Albert (1985), *Measuring the Universe: Cosmic Dimensions from Aristarchus to Halley*, University of Chicago Press, Chicago and London.

Wilson, Curtis (1989), *Astronomy from Kepler to Newton*, Variorum, Aldershot.

# Part III

# Miscellaneous

# Some Interesting Variable Stars

JOHN ISLES

All variable stars are of potential interest, and hundreds of them can be observed with the slightest optical aid – even with a pair of binoculars. The stars in the list that follows include many that are popular with amateur observers, as well as some less well-known objects that are nevertheless suitable for study visually. The periods and ranges of many variables are not constant from one cycle to another, and some are completely irregular.

Finder charts are given after the list for those stars marked with an asterisk. These charts are adapted with permission from those issued by the Variable Star Section of the British Astronomical Association. Apart from the eclipsing variables and others in which the light changes are purely a geometrical effect, variable stars can be divided broadly into two classes: the pulsating stars, and the eruptive or cataclysmic variables.

Mira (Omicron Ceti) is the best-known member of the long-period subclass of pulsating red-giant stars. The chart is suitable for use in estimating the magnitude of Mira when it reaches naked-eye brightness – typically from about a month before the predicted date of maximum until two or three months after maximum. Predictions for Mira and other stars of its class follow the section of finder charts.

The semi-regular variables are less predictable, and generally have smaller ranges. V Canum Venaticorum is one of the more reliable ones, with steady oscillations in a six-month cycle. Z Ursae Majoris, easily found with binoculars near Delta, has a large range, and often shows double maxima owing to the presence of multiple periodicities in its light changes. The chart for Z is also suitable for observing another semi-regular star, RY Ursae Majoris. These semi-regular stars are mostly red giants or supergiants.

The RV Tauri stars are of earlier spectral class than the semi-regulars, and in a full cycle of variation they often show deep minima and double maxima that are separated by a secondary minimum. U Monocerotis is one of the brightest RV Tauri stars.

Among eruptive variable stars is the carbon-rich supergiant R Coronae Borealis. Its unpredictable eruptions cause it not to brighten, but to fade. This happens when one of the sooty clouds that the star throws out from time to time happens to come in our direction and blots out most of the star's light from our view. Much of the time R Coronae is bright enough to be seen with binoculars, and the chart can be used to estimate its magnitude. During the deepest minima, however, the star needs a telescope of 25 centimetres or larger aperture to be detected.

CH Cygni is a symbiotic star – that is, a close binary comprising a red giant and a hot dwarf star that interact physically, giving rise to outbursts. The system also shows semi-regular oscillations, and sudden fades and rises that may be connected with eclipses.

Observers can follow the changes of these variable stars by using the comparison stars whose magnitudes are given below each chart. Observations of variable stars by amateurs are of scientific value, provided they are collected and made available for analysis. This is done by several organizations, including the British Astronomical Association (see the list of astronomical societies in this volume), the American Association of Variable Star Observers (25 Birch Street, Cambridge, Mass. 02138), and the Royal Astronomical Society of New Zealand (P.O. Box 3181, Wellington).

| Star | RA | | Declination | | Range | Type | Period | Spectrum |
|------|----|----|-------------|----|-------|------|--------|----------|
| | h | m | ° | ′ | | | (days) | |
| R Andromedae | 00 | 24.0 | +38 | 35 | 5.8–14.9 | Mira | 409 | S |
| W Andromedae | 02 | 17.6 | +44 | 18 | 6.7–14.6 | Mira | 396 | S |
| U Antliae | 10 | 35.2 | −39 | 34 | 5–6 | Irregular | — | C |
| Theta Apodis | 14 | 05.3 | −76 | 48 | 5–7 | Semi-regular | 119 | M |
| R Aquarii | 23 | 43.8 | −15 | 17 | 5.8–12.4 | Symbiotic | 387 | M+Pec |
| T Aquarii | 20 | 49.9 | −05 | 09 | 7.2–14.2 | Mira | 202 | M |
| R Aquilae | 19 | 06.4 | +08 | 14 | 5.5–12.0 | Mira | 284 | M |
| V Aquilae | 19 | 04.4 | −05 | 41 | 6.6–8.4 | Semi-regular | 353 | C |
| Eta Aquilae | 19 | 52.5 | +01 | 00 | 3.5–4.4 | Cepheid | 7.2 | F–G |
| U Arae | 17 | 53.6 | −51 | 41 | 7.7–14.1 | Mira | 225 | M |
| R Arietis | 02 | 16.1 | +25 | 03 | 7.4–13.7 | Mira | 187 | M |
| U Arietis | 03 | 11.0 | +14 | 48 | 7.2–15.2 | Mira | 371 | M |
| R Aurigae | 05 | 17.3 | +53 | 35 | 6.7–13.9 | Mira | 458 | M |
| Epsilon Aurigae | 05 | 02.0 | +43 | 49 | 2.9–3.8 | Algol | 9892 | F+B |
| R Boötis | 14 | 37.2 | +26 | 44 | 6.2–13.1 | Mira | 223 | M |

| Star | RA | | Declination | | Range | Type | Period | Spectrum |
|------|----|----|----|----|-------|------|--------|----------|
| | h | m | ° | ' | | | (days) | |
| X Camelopardalis | 04 | 45.7 | +75 | 06 | 7.4–14.2 | Mira | 144 | K–M |
| R Cancri | 08 | 16.6 | +11 | 44 | 6.1–11.8 | Mira | 362 | M |
| X Cancri | 08 | 55.4 | +17 | 14 | 5.6–7.5 | Semi-regular | 195? | C |
| R Canis Majoris | 07 | 19.5 | −16 | 24 | 5.7–6.3 | Algol | 1.1 | F |
| VY Canis Majoris | 07 | 23.0 | −25 | 46 | 6.5–9.6 | Unique | — | M |
| S Canis Minoris | 07 | 32.7 | +08 | 19 | 6.6–13.2 | Mira | 333 | M |
| R Canum Ven. | 13 | 49.0 | +39 | 33 | 6.5–12.9 | Mira | 329 | M |
| *V Canum Ven. | 13 | 19.5 | +45 | 32 | 6.5–8.6 | Semi-regular | 192 | M |
| R Carinae | 09 | 32.2 | −62 | 47 | 3.9–10.5 | Mira | 309 | M |
| S Carinae | 10 | 09.4 | −61 | 33 | 4.5–9.9 | Mira | 149 | K–M |
| I Carinae | 09 | 45.2 | −62 | 30 | 3.3–4.2 | Cepheid | 35.5 | F–K |
| Eta Carinae | 10 | 45.1 | −59 | 41 | −0.8–7.9 | Irregular | — | Pec |
| R Cassiopeiae | 23 | 58.4 | +51 | 24 | 4.7–13.5 | Mira | 430 | M |
| S Cassiopeiae | 01 | 19.7 | +72 | 37 | 7.9–16.1 | Mira | 612 | S |
| W Cassiopeiae | 00 | 54.9 | +58 | 34 | 7.8–12.5 | Mira | 406 | C |
| Gamma Cas. | 00 | 56.7 | +60 | 43 | 1.6–3.0 | Gamma Cas. | — | B |
| Rho Cassiopeiae | 23 | 54.4 | +57 | 30 | 4.1–6.2 | Semi-regular | — | F–K |
| R Centauri | 14 | 16.6 | −59 | 55 | 5.3–11.8 | Mira | 546 | M |
| S Centauri | 12 | 24.6 | −49 | 26 | 7–8 | Semi-regular | 65 | C |
| T Centauri | 13 | 41.8 | −33 | 36 | 5.5–9.0 | Semi-regular | 90 | K–M |
| S Cephei | 21 | 35.2 | +78 | 37 | 7.4–12.9 | Mira | 487 | C |
| T Cephei | 21 | 09.5 | +68 | 29 | 5.2–11.3 | Mira | 388 | M |
| Delta Cephei | 22 | 29.2 | +58 | 25 | 3.5–4.4 | Cepheid | 5.4 | F–G |
| Mu Cephei | 21 | 43.5 | +58 | 47 | 3.4–5.1 | Semi-regular | 730 | M |
| U Ceti | 02 | 33.7 | −13 | 09 | 6.8–13.4 | Mira | 235 | M |
| W Ceti | 00 | 02.1 | −14 | 41 | 7.1–14.8 | Mira | 351 | S |
| *Omicron Ceti | 02 | 19.3 | −02 | 59 | 2.0–10.1 | Mira | 332 | M |
| R Chamaeleontis | 08 | 21.8 | −76 | 21 | 7.5–14.2 | Mira | 335 | M |
| T Columbae | 05 | 19.3 | −33 | 42 | 6.6–12.7 | Mira | 226 | M |
| R Comae Ber. | 12 | 04.3 | +18 | 47 | 7.1–14.6 | Mira | 363 | M |
| *R Coronae Bor. | 15 | 48.6 | +28 | 09 | 5.7–14.8 | R Coronae Bor. | — | C |
| S Coronae Bor. | 15 | 21.4 | +31 | 22 | 5.8–14.1 | Mira | 360 | M |
| T Coronae Bor. | 15 | 59.6 | +25 | 55 | 2.0–10.8 | Recurrent nova | — | M+Pec |
| V Coronae Bor. | 15 | 49.5 | +39 | 34 | 6.9–12.6 | Mira | 358 | C |
| W Coronae Bor. | 16 | 15.4 | +37 | 48 | 7.8–14.3 | Mira | 238 | M |
| R Corvi | 12 | 19.6 | −19 | 15 | 6.7–14.4 | Mira | 317 | M |
| R Crucis | 12 | 23.6 | −61 | 38 | 6.4–7.2 | Cepheid | 5.8 | F–G |
| R Cygni | 19 | 36.8 | +50 | 12 | 6.1–14.4 | Mira | 426 | S |
| U Cygni | 20 | 19.6 | +47 | 54 | 5.9–12.1 | Mira | 463 | C |
| W Cygni | 21 | 36.0 | +45 | 22 | 5.0–7.6 | Semi-regular | 131 | M |

| Star | RA | | Declination | | Range | Type | Period | Spectrum |
|------|----|----|----|----|----|----|----|----|
| | h | m | ° | ′ | | | (days) | |
| RT Cygni | 19 | 43.6 | +48 | 47 | 6.0−13.1 | Mira | 190 | M |
| SS Cygni | 21 | 42.7 | +43 | 35 | 7.7−12.4 | Dwarf nova | 50± | K+Pec |
| *CH Cygni | 19 | 24.5 | +50 | 14 | 5.6−9.0 | Symbiotic | — | M+B |
| Chi Cygni | 19 | 50.6 | +32 | 55 | 3.3−14.2 | Mira | 408 | S |
| R Delphini | 20 | 14.9 | +09 | 05 | 7.6−13.8 | Mira | 285 | M |
| U Delphini | 20 | 45.5 | +18 | 05 | 5.6−7.5 | Semi-regular | 110? | M |
| EU Delphini | 20 | 37.9 | +18 | 16 | 5.8−6.9 | Semi-regular | 60 | M |
| Beta Doradûs | 05 | 33.6 | −62 | 29 | 3.5−4.1 | Cepheid | 9.8 | F−G |
| R Draconis | 16 | 32.7 | +66 | 45 | 6.7−13.2 | Mira | 246 | M |
| T Eridani | 03 | 55.2 | −24 | 02 | 7.2−13.2 | Mira | 252 | M |
| R Fornacis | 02 | 29.3 | −26 | 06 | 7.5−13.0 | Mira | 389 | C |
| R Geminorum | 07 | 07.4 | +22 | 42 | 6.0−14.0 | Mira | 370 | S |
| U Geminorum | 07 | 55.1 | +22 | 00 | 8.2−14.9 | Dwarf nova | 105± | Pec+M |
| Zeta Geminorum | 07 | 04.1 | +20 | 34 | 3.6−4.2 | Cepheid | 10.2 | F−G |
| Eta Geminorum | 06 | 14.9 | +22 | 30 | 3.2−3.9 | Semi-regular | 233 | M |
| S Gruis | 22 | 26.1 | −48 | 26 | 6.0−15.0 | Mira | 402 | M |
| S Herculis | 16 | 51.9 | +14 | 56 | 6.4−13.8 | Mira | 307 | M |
| U Herculis | 16 | 25.8 | +18 | 54 | 6.4−13.4 | Mira | 406 | M |
| Alpha Herculis | 17 | 14.6 | +14 | 23 | 2.7−4.0 | Semi-regular | — | M |
| 68, u Herculis | 17 | 17.3 | +33 | 06 | 4.7−5.4 | Algol | 2.1 | B+B |
| R Horologii | 02 | 53.9 | −49 | 53 | 4.7−14.3 | Mira | 408 | M |
| U Horologii | 03 | 52.8 | −45 | 50 | 6−14 | Mira | 348 | M |
| R Hydrae | 13 | 29.7 | −23 | 17 | 3.5−10.9 | Mira | 389 | M |
| U Hydrae | 10 | 37.6 | −13 | 23 | 4.3−6.5 | Semi-regular | 450? | C |
| VW Hydri | 04 | 09.1 | −71 | 18 | 8.4−14.4 | Dwarf nova | 27± | Pec |
| R Leonis | 09 | 47.6 | +11 | 26 | 4.4−11.3 | Mira | 310 | M |
| R Leonis Minoris | 09 | 45.6 | +34 | 31 | 6.3−13.2 | Mira | 372 | M |
| R Leporis | 04 | 59.6 | −14 | 48 | 5.5−11.7 | Mira | 427 | C |
| Y Librae | 15 | 11.7 | −06 | 01 | 7.6−14.7 | Mira | 276 | M |
| RS Librae | 15 | 24.3 | −22 | 55 | 7.0−13.0 | Mira | 218 | M |
| Delta Librae | 15 | 01.0 | −08 | 31 | 4.9−5.9 | Algol | 2.3 | A |
| R Lyncis | 07 | 01.3 | +55 | 20 | 7.2−14.3 | Mira | 379 | S |
| R Lyrae | 18 | 55.3 | +43 | 57 | 3.9−5.0 | Semi-regular | 46? | M |
| RR Lyrae | 19 | 25.5 | +42 | 47 | 7.1−8.1 | RR Lyrae | 0.6 | A−F |
| Beta Lyrae | 18 | 50.1 | +33 | 22 | 3.3−4.4 | Eclipsing | 12.9 | B |
| U Microscopii | 20 | 29.2 | −40 | 25 | 7.0−14.4 | Mira | 334 | M |
| *U Monocerotis | 07 | 30.8 | −09 | 47 | 5.9−7.8 | RV Tauri | 91 | F−K |
| V Monocerotis | 06 | 22.7 | −02 | 12 | 6.0−13.9 | Mira | 340 | M |
| R Normae | 15 | 36.0 | −49 | 30 | 6.5−13.9 | Mira | 508 | M |
| T Normae | 15 | 44.1 | −54 | 59 | 6.2−13.6 | Mira | 241 | M |

| Star | RA | | Declination | | Range | Type | Period | Spectrum |
|---|---|---|---|---|---|---|---|---|
| | h | m | ° | ′ | | | (days) | |
| R Octantis | 05 | 26.1 | −86 | 23 | 6.3−13.2 | Mira | 405 | M |
| S Octantis | 18 | 08.7 | −86 | 48 | 7.2−14.0 | Mira | 259 | M |
| V Ophiuchi | 16 | 26.7 | −12 | 26 | 7.3−11.6 | Mira | 297 | C |
| X Ophiuchi | 18 | 38.3 | +08 | 50 | 5.9−9.2 | Mira | 329 | M |
| RS Ophiuchi | 17 | 50.2 | −06 | 43 | 4.3−12.5 | Recurrent nova | — | OB+M |
| U Orionis | 05 | 55.8 | +20 | 10 | 4.8−13.0 | Mira | 368 | M |
| W Orionis | 05 | 05.4 | +01 | 11 | 5.9−7.7 | Semi-regular | 212 | C |
| Alpha Orionis | 05 | 55.2 | +07 | 24 | 0.0−1.3 | Semi-regular | 2335 | M |
| S Pavonis | 19 | 55.2 | −59 | 12 | 6.6−10.4 | Semi-regular | 381 | M |
| Kappa Pavonis | 18 | 56.9 | −67 | 14 | 3.9−4.8 | W Virginis | 9.1 | G |
| R Pegasi | 23 | 06.8 | +10 | 33 | 6.9−13.8 | Mira | 378 | M |
| X Persei | 03 | 55.4 | +31 | 03 | 6.0−7.0 | Gamma Cas. | — | O9.5 |
| Beta Persei | 03 | 08.2 | +40 | 57 | 2.1−3.4 | Algol | 2.9 | B |
| Zeta Phoenicis | 01 | 08.4 | −55 | 15 | 3.9−4.4 | Algol | 1.7 | B+B |
| R Pictoris | 04 | 46.2 | −49 | 15 | 6.4−10.1 | Semi-regular | 171 | M |
| RS Puppis | 08 | 13.1 | −34 | 35 | 6.5−7.7 | Cepheid | 41.4 | F−G |
| L$^2$ Puppis | 07 | 13.5 | −44 | 39 | 2.6−6.2 | Semi-regular | 141 | M |
| T Pyxidis | 09 | 04.7 | −32 | 23 | 6.5−15.3 | Recurrent nova | 7000± | Pec |
| U Sagittae | 19 | 18.8 | +19 | 37 | 6.5−9.3 | Algol | 3.4 | B+G |
| WZ Sagittae | 20 | 07.6 | +17 | 42 | 7.0−15.5 | Dwarf nova | 1900± | A |
| R Sagittarii | 19 | 16.7 | −19 | 18 | 6.7−12.8 | Mira | 270 | M |
| RR Sagittarii | 19 | 55.9 | −29 | 11 | 5.4−14.0 | Mira | 336 | M |
| RT Sagittarii | 20 | 17.7 | −39 | 07 | 6.0−14.1 | Mira | 306 | M |
| RU Sagittarii | 19 | 58.7 | −41 | 51 | 6.0−13.8 | Mira | 240 | M |
| RY Sagittarii | 19 | 16.5 | −33 | 31 | 5.8−14.0 | R Coronae Bor. | — | G |
| RR Scorpii | 16 | 56.6 | −30 | 35 | 5.0−12.4 | Mira | 281 | M |
| RS Scorpii | 16 | 55.6 | −45 | 06 | 6.2−13.0 | Mira | 320 | M |
| RT Scorpii | 17 | 03.5 | −36 | 55 | 7.0−15.2 | Mira | 449 | S |
| Delta Scorpii | 16 | 00.3 | −22 | 37 | 1.6−2.3 | Irregular | — | B |
| S Sculptoris | 00 | 15.4 | −32 | 03 | 5.5−13.6 | Mira | 363 | M |
| R Scuti | 18 | 47.5 | −05 | 42 | 4.2−8.6 | RV Tauri | 146 | G−K |
| R Serpentis | 15 | 50.7 | +15 | 08 | 5.2−14.4 | Mira | 356 | M |
| S Serpentis | 15 | 21.7 | +14 | 19 | 7.0−14.1 | Mira | 372 | M |
| T Tauri | 04 | 22.0 | +19 | 32 | 9.3−13.5 | T Tauri | — | F−K |
| SU Tauri | 05 | 49.1 | +19 | 04 | 9.1−16.9 | R Coronae Bor. | — | G |
| Lambda Tauri | 04 | 00.7 | +12 | 29 | 3.4−3.9 | Algol | 4.0 | B+A |
| R Trianguli | 02 | 37.0 | +34 | 16 | 5.4−12.6 | Mira | 267 | M |
| R Ursae Majoris | 10 | 44.6 | +68 | 47 | 6.5−13.7 | Mira | 302 | M |
| T Ursae Majoris | 12 | 36.4 | +59 | 29 | 6.6−13.5 | Mira | 257 | M |
| *Z Ursae Majoris | 11 | 56.5 | +57 | 52 | 6.2−9.4 | Semi-regular | 196 | M |

| Star | RA | | Declination | | Range | Type | Period (days) | Spectrum |
|---|---|---|---|---|---|---|---|---|
| | h | m | ° | ′ | | | | |
| *RY Ursae Majoris | 12 | 20.5 | +61 | 19 | 6.7−8.3 | Semi-regular | 310? | M |
| U Ursae Minoris | 14 | 17.3 | +66 | 48 | 7.1−13.0 | Mira | 331 | M |
| R Virginis | 12 | 38.5 | +06 | 59 | 6.1−12.1 | Mira | 146 | M |
| S Virginis | 13 | 33.0 | −07 | 12 | 6.3−13.2 | Mira | 375 | M |
| SS Virginis | 12 | 25.3 | +00 | 48 | 6.0−9.6 | Semi-regular | 364 | C |
| R Vulpeculae | 21 | 04.4 | +23 | 49 | 7.0−14.3 | Mira | 137 | M |
| Z Vulpeculae | 19 | 21.7 | +25 | 34 | 7.3−8.9 | Algol | 2.5 | B+A |

**V CANUM VENATICORUM**      13h 19.5m +45° 32′ (2000)

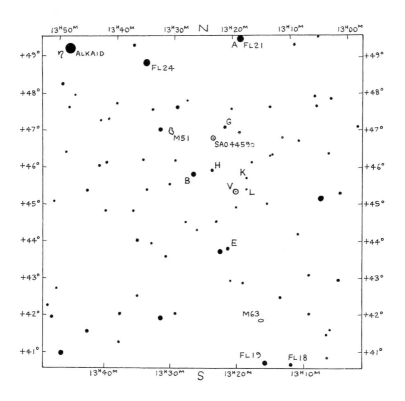

| | |
|---|---|
| A 5.1 | H 7.8 |
| B 5.9 | K 8.4 |
| E 6.5 | L 8.6 |
| G 7.1 | |

**o (MIRA) CETI    02h 19.3m −02° 59′ (2000)**

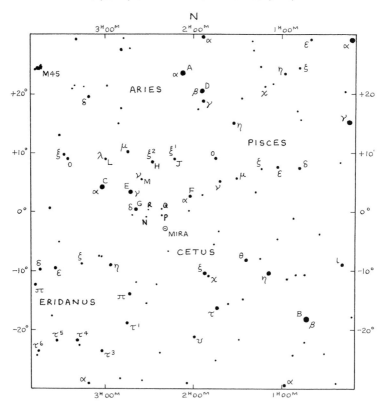

| | | | |
|---|---|---|---|
| A | 2.2 | J | 4.4 |
| B | 2.4 | L | 4.9 |
| C | 2.7 | M | 5.1 |
| D | 3.0 | N | 5.4 |
| E | 3.6 | P | 5.5 |
| F | 3.8 | Q | 5.7 |
| G | 4.1 | R | 6.1 |
| H | 4.3 | | |

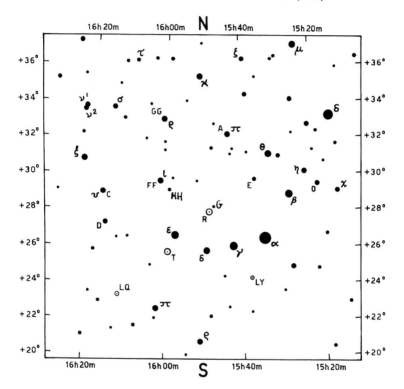

R CORONAE BOREALIS    15h 48.6m +28° 09′ (2000)

| | |
|---|---|
| FF 5.0 | C 5.8 |
| GG 5.4 | D 6.2 |
| A 5.6 | E 6.5 |
| HH 7.1 | |
| G 7.4 | |

## CH CYGNI    19h 24.5m +50° 14′ (2000)

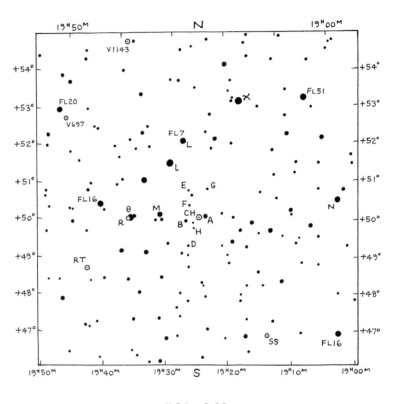

| N 5.4 | D 8.0 |
|-------|-------|
| M 5.5 | E 8.1 |
| L 5.8 | F 8.5 |
| A 6.5 | G 8.5 |
| B 7.4 | H 9.2 |

## U MONOCEROTIS    07h 30.8m −09° 47′ (2000)

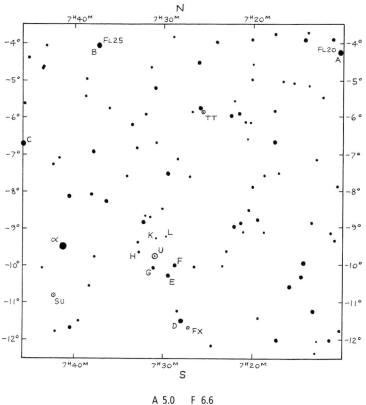

| A 5.0 | F 6.6 |
|-------|-------|
| B 5.2 | G 7.0 |
| C 5.7 | H 7.5 |
| D 5.9 | K 7.8 |
| E 6.0 | L 8.0 |

**RY URSAE MAJORIS    12h 20.5m +61° 19′ (2000)**
**Z URSAE MAJORIS     11h 56.5m +57° 52′ (2000)**

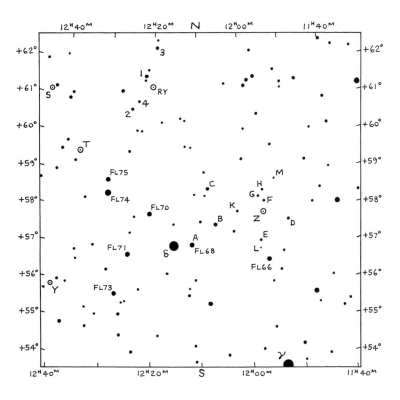

| A 6.5 | F 8.6 | M 9.1 |
|-------|-------|-------|
| B 7.2 | G 8.7 | 1 6.9 |
| C 7.6 | H 8.8 | 2 7.4 |
| D 8.0 | K 8.9 | 3 7.7 |
| E 8.3 | L 9.0 | 4 7.8 |

# Mira Stars: Maxima, 2005

## JOHN ISLES

Below are the predicted dates of maxima for Mira stars that reach magnitude 7.5 or brighter at an average maximum. Individual maxima can in some cases be brighter or fainter than average by a magnitude or more, and all dates are only approximate. The positions, extreme ranges and mean periods of these stars can be found in the preceding list of interesting variable stars.

| Star | Mean Magnitude at Maximum | Dates of Maxima |
|---|---|---|
| R Andromedae | 6.9 | 15 Feb |
| W Andromedae | 7.4 | 8 Feb |
| R Aquarii | 6.5 | 6 Oct |
| R Aquilae | 6.1 | 2 Apr |
| R Bootis | 7.2 | 18 Apr, 28 Nov |
| R Cancri | 6.8 | 9 Oct |
| S Canis Minoris | 7.5 | 21 May |
| R Carinae | 4.6 | 7 Jan, 12 Nov |
| S Carinae | 5.7 | 26 Apr, 23 Sep |
| R Cassiopeiae | 7.0 | 29 May |
| R Centauri | 5.8 | 7 Sep |
| T Cephei | 6.0 | 5 Nov |
| U Ceti | 7.5 | 8 Feb, 1 Oct |
| Omicron Ceti | 3.4 | 5 May |
| T Columbae | 7.5 | 18 May, 30 Dec |
| S Coronae Borealis | 7.3 | 6 Oct |
| V Coronae Borealis | 7.5 | 27 Mar |
| R Corvi | 7.5 | 25 Jul |
| R Cygni | 7.5 | 1 Jul |
| U Cygni | 7.2 | 23 Jan |
| RT Cygni | 7.3 | 10 Jun, 18 Dec |
| chi Cygni | 5.2 | 4 Jul |

| Star | Mean Magnitude at Maximum | Dates of Maxima |
|------|---------------------------|-----------------|
| R Geminorum | 7.1 | 2 Dec |
| U Herculis | 7.5 | 18 Jun |
| R Horologii | 6.0 | 15 Aug |
| R Hydrae | 4.5 | 22 Sep |
| R Leonis | 5.8 | 6 Jun |
| R Leonis Minoris | 7.1 | 26 Nov |
| R Leporis | 6.8 | 5 Mar |
| RS Librae | 7.5 | 8 May, 11 Dec |
| V Monocerotis | 7.0 | 8 Oct |
| T Normae | 7.4 | 11 Aug |
| V Ophiuchi | 7.5 | 13 Feb, 7 Dec |
| X Ophiuchi | 6.8 | 5 Apr |
| U Orionis | 6.3 | 1 Jan |
| R Sagittarii | 7.3 | 13 Jul |
| RR Sagittarii | 6.8 | 16 May |
| RT Sagittarii | 7.0 | 10 Apr |
| RU Sagittarii | 7.2 | 2 Mar, 29 Oct |
| RR Scorpii | 5.9 | 29 Apr |
| RS Scorpii | 7.0 | 1 Oct |
| S Sculptoris | 6.7 | 21 Dec |
| R Serpentis | 6.9 | 17 Nov |
| R Trianguli | 6.2 | 23 Apr |
| R Ursae Majoris | 7.5 | 16 Jan, 15 Dec |
| R Virginis | 6.9 | 25 Mar, 17 Aug |

# Some Interesting Double Stars

## BOB ARGYLE

The positions, angles and separations given below correspond to epoch 2005.0.

| No. | RA | | Declin-ation | | Star | Magni-tudes | Separa-tion | PA | Cata-logue | Comments |
|---|---|---|---|---|---|---|---|---|---|---|
| | h | m | ° | ' | | | arcsec | ° | | |
| 1 | 00 | 31.5 | −62 | 58 | β Tuc | 4.4, 4.8 | 27.1 | 169 | LCL 119 | Both again difficult doubles. |
| 2 | 00 | 49.1 | +57 | 49 | η Cas | 3.4, 7.5 | 13.0 | 319 | Σ60 | Easy. Creamy, bluish. |
| 3 | 00 | 55.0 | +23 | 38 | 36 And | 6.0, 6.4 | 1.0 | 317 | Σ73 | P = 168 years. Both yellow. Slowly opening. |
| 4 | 01 | 13.7 | +07 | 35 | ζ Psc | 5.6, 6.5 | 23.1 | 63 | Σ100 | Yellow, reddish-white. |
| 5 | 01 | 39.8 | −56 | 12 | p Eri | 5.8, 5.8 | 11.6 | 189 | Δ5 | Period = 483 years. |
| 6 | 01 | 53.5 | +19 | 18 | γ Ari | 4.8, 4.8 | 7.5 | 1 | Σ180 | Very easy. Both white. |
| 7 | 02 | 02.0 | +02 | 46 | α Psc | 4.2, 5.1 | 1.8 | 268 | Σ202 | Binary, period = 933 years. |
| 8 | 02 | 03.9 | +42 | 20 | γ And | 2.3, 5.0 | 9.6 | 63 | Σ205 | Yellow, blue. Relatively fixed. |
| | | | | | γ2 And | 5.1, 6.3 | 0.3 | 101 | OΣ38 | BC. Needs 30 cm. Closing. |
| 9 | 02 | 29.1 | +67 | 24 | ι Cas AB | 4.9, 6.9 | 2.6 | 230 | Σ262 | AB is long-period binary. P = 620 years. |
| | | | | | ι Cas AC | 4.9, 8.4 | 7.2 | 118 | | |
| 10 | 02 | 33.8 | −28 | 14 | ω For | 5.0, 7.7 | 10.8 | 245 | HJ 3506 | Common proper motion. |
| 11 | 02 | 43.3 | +03 | 14 | γ Cet | 3.5, 7.3 | 2.6 | 298 | Σ299 | Not too easy. |

| No. | RA | | Declin-ation | | Star | Magni-tudes | Separa-tion | PA | Cata-logue | Comments |
|-----|-----|-----|-----|-----|-----|-----|-----|-----|-----|-----|
| | h | m | ° | ′ | | | arcsec | ° | | |
| 12 | 02 | 58.3 | −40 | 18 | θ Eri | 3.4, 4.5 | 8.3 | 90 | PZ 2 | Both white. |
| 13 | 02 | 59.2 | +21 | 20 | ε Ari | 5.2, 5.5 | 1.5 | 208 | Σ333 | Binary. Little motion. Both white. |
| 14 | 03 | 00.9 | +52 | 21 | Σ331 Per | 5.3, 6.7 | 12.0 | 85 | – | Fixed. |
| 15 | 03 | 12.1 | −28 | 59 | α For | 4.0, 7.0 | 5.1 | 299 | HJ 3555 | P = 269 years. B variable? |
| 16 | 03 | 48.6 | −37 | 37 | f Eri | 4.8, 5.3 | 8.2 | 215 | Δ16 | Pale yellow. Fixed. |
| 17 | 03 | 54.3 | −02 | 57 | 32 Eri | 4.8, 6.1 | 6.9 | 348 | Σ470 | Fixed. |
| 18 | 04 | 32.0 | +53 | 55 | 1 Cam | 5.7, 6.8 | 10.3 | 308 | Σ550 | Fixed. |
| 19 | 04 | 50.9 | −53 | 28 | ι Pic | 5.6, 6.4 | 12.4 | 58 | Δ18 | Good object for small apertures. Fixed. |
| 20 | 05 | 13.2 | −12 | 56 | κ Lep | 4.5, 7.4 | 2.2 | 357 | Σ661 | Visible in 7.5 cm. |
| 21 | 05 | 14.5 | −08 | 12 | β Ori | 0.1, 6.8 | 9.5 | 204 | Σ668 | Companion once thought to be close double. |
| 22 | 05 | 21.8 | −24 | 46 | 41 Lep | 5.4, 6.6 | 3.4 | 93 | HJ 3752 | Deep-yellow pair in a rich field. |
| 23 | 05 | 24.5 | −02 | 24 | η Ori | 3.8, 4.8 | 1.7 | 78 | DA 5 | Slow-moving binary. |
| 24 | 05 | 35.1 | +09 | 56 | λ Ori | 3.6, 5.5 | 4.3 | 44 | Σ738 | Fixed. |
| 25 | 05 | 35.3 | −05 | 23 | θ Ori AB | 6.7, 7.9 | 8.6 | 32 | Σ748 | Trapezium in M42. |
| | | | | | θ Ori CD | 5.1, 6.7 | 13.4 | 61 | | |
| 26 | 05 | 38.7 | −02 | 36 | σ Ori AC | 4.0, 10.3 | 11.4 | 238 | Σ762 | Quintuple. A is a close double. |
| | | | | | σ Ori ED | 6.5, 7.5 | 30.1 | 231 | | |
| 27 | 05 | 40.7 | −01 | 57 | ζ Ori | 1.9, 4.0 | 2.4 | 164 | Σ774 | Can be split in 7.5 cm. Long-period binary. |
| 28 | 06 | 14.9 | +22 | 30 | η Gem | var, 6.5 | 1.6 | 255 | β1008 | Well seen with 20 cm. Primary orange. |
| 29 | 06 | 46.2 | +59 | 27 | 12 Lyn AB | 5.4, 6.0 | 1.7 | 68 | Σ948 | AB is binary, P = 706 years. |
| | | | | | 12 Lyn AC | 5.4, 7.3 | 8.7 | 309 | | |

| No. | RA | | Declination | | Star | Magnitudes | Separation | PA | Catalogue | Comments |
|-----|----|----|-------------|---|------|------------|------------|----|-----------|----------|
|     | h  | m  | °           | ′ |      |            | arcsec     | °  |           |          |
| 30 | 07 | 08.7 | −70 | 30 | γ Vol | 3.9, 5.8 | 14.1 | 298 | Δ42 | Very slow binary. |
| 31 | 07 | 16.6 | −23 | 19 | h3945 CMa | 4.8, 6.8 | 26.8 | 51 | − | Contrasting colours. |
| 32 | 07 | 20.1 | +21 | 59 | δ Gem | 3.5, 8.2 | 5.7 | 226 | Σ1066 | Not too easy. Yellow, pale blue. |
| 33 | 07 | 34.6 | +31 | 53 | α Gem | 1.9, 2.9 | 4.3 | 61 | Σ1110 | Widening. Easy with 7.5 cm. |
| 34 | 07 | 38.8 | −26 | 48 | κ Pup | 4.5, 4.7 | 9.8 | 318 | H III 27 | Both white. |
| 35 | 08 | 12.2 | +17 | 39 | ζ Cnc AB | 5.6, 6.0 | 1.0 | 57 | Σ1196 | Period (AB) = 60 years. Near maximum separation. |
|    |    |      |     |    | ζ Cnc AB-C | 5.0, 6.2 | 5.9 | 71 | Σ1196 | Period (AB-C) = 1,150 years. |
| 36 | 08 | 44.7 | −54 | 43 | δ Vel | 2.1, 5.1 | 0.9 | 331 | I 10 | Difficult close pair. Period 142 years. |
| 37 | 08 | 46.8 | +06 | 25 | ε Hyd | 3.3, 6.8 | 2.9 | 301 | Σ1273 | PA slowly increasing. A is a very close pair. |
| 38 | 09 | 18.8 | +36 | 48 | 38 Lyn | 3.9, 6.6 | 2.8 | 230 | Σ1338 | Almost fixed. |
| 39 | 09 | 47.1 | −65 | 04 | μ Car | 3.1, 6.1 | 5.0 | 128 | RMK 11 | Fixed. Fine in small telescopes. |
| 40 | 10 | 20.0 | +19 | 50 | γ Leo | 2.2, 3.5 | 4.4 | 125 | Σ1424 | Binary, period = 619 years. Both orange. |
| 41 | 10 | 32.0 | −45 | 04 | s Vel | 6.2, 6.5 | 13.5 | 218 | PZ 3 | Fixed. |
| 42 | 10 | 46.8 | −49 | 26 | μ Vel | 2.7, 6.4 | 2.5 | 54 | R 155 | P = 138 years. Near widest separation. |
| 43 | 10 | 55.6 | +24 | 45 | 54 Leo | 4.5, 6.3 | 6.6 | 111 | Σ1487 | Slowly widening. Pale yellow and white. |

| No. | RA | | Declin-ation | | Star | Magni-tudes | Separa-tion | PA | Cata-logue | Comments |
|---|---|---|---|---|---|---|---|---|---|---|
| | h | m | ° | ′ | | | arcsec | ° | | |
| 44 | 11 | 18.2 | +31 | 32 | ξ UMa | 4.3, 4.8 | 1.7 | 246 | Σ1523 | Binary, 60 years. Needs 7.5 cm. |
| 45 | 11 | 21.0 | −54 | 29 | π Cen | 4.3, 5.0 | 0.2 | 156 | I 879 | Binary, 38.7 years. Very close. Needs 35 cm. |
| 46 | 11 | 23.9 | +10 | 32 | ι Leo | 4.0, 6.7 | 1.8 | 105 | Σ1536 | Binary, period = 186 years. Slowly widening. |
| 47 | 11 | 32.3 | −29 | 16 | N Hya | 5.8, 5.9 | 9.5 | 210 | H III 96 | Fixed. |
| 48 | 12 | 14.0 | −45 | 43 | D Cen | 5.6, 6.8 | 2.8 | 243 | RMK 14 | Orange and white. Closing. |
| 49 | 12 | 26.6 | −63 | 06 | α Cru | 1.4, 1.9 | 4.0 | 112 | Δ252 | Third star in a low-power field. |
| 50 | 12 | 41.5 | −48 | 58 | γ Cen | 2.9, 2.9 | 0.7 | 341 | HJ 4539 | Period = 84 years. Closing. Both yellow. |
| 51 | 12 | 41.7 | −01 | 27 | γ Vir | 3.5, 3.5 | 0.4 | 168 | Σ1670 | Rapid motion near periastron. See page 201. |
| 52 | 12 | 46.3 | −68 | 06 | β Mus | 3.7, 4.0 | 1.3 | 46 | R 207 | Both white. Closing slowly. P = 383 years. |
| 53 | 12 | 54.6 | −57 | 11 | μ Cru | 4.3, 5.3 | 34.9 | 17 | Δ126 | Fixed. Both white. |
| 54 | 12 | 56.0 | +38 | 19 | α CVn | 2.9, 5.5 | 19.3 | 229 | Σ1692 | Easy. Yellow, bluish. |
| 55 | 13 | 22.6 | −60 | 59 | J Cen | 4.6, 6.5 | 60.0 | 343 | Δ133 | Fixed. A is a close pair. |
| 56 | 13 | 24.0 | +54 | 56 | ζ UMa | 2.3, 4.0 | 14.4 | 152 | Σ1744 | Very easy. Naked-eye pair with Alcor. |
| 57 | 13 | 51.8 | −33 | 00 | 3 Cen | 4.5, 6.0 | 7.9 | 106 | H III 101 | Both white. Closing slowly. |
| 58 | 14 | 39.6 | −60 | 50 | α Cen | 0.0, 1.2 | 10.5 | 230 | RHD 1 | Finest pair in the sky. P = 80 years. Closing. |

| No. | RA | | Declin-ation | | Star | Magni-tudes | Separa-tion | PA | Cata-logue | Comments |
|---|---|---|---|---|---|---|---|---|---|---|
| | h | m | ° | ' | | | arcsec | ° | | |
| 59 | 14 | 41.1 | +13 | 44 | ζ Boo | 4.5, 4.6 | 0.7 | 297 | Σ1865 | Both white. Closing – highly inclined orbit. |
| 60 | 14 | 45.0 | +27 | 04 | ε Boo | 2.5, 4.9 | 2.9 | 345 | Σ1877 | Yellow, blue. Fine pair. |
| 61 | 14 | 46.0 | −25 | 27 | 54 Hya | 5.1, 7.1 | 8.3 | 122 | H III 97 | Closing slowly. |
| 62 | 14 | 49.3 | −14 | 09 | μ Lib | 5.8, 6.7 | 1.9 | 2 | β106 | Becoming wider. Fine in 7.5 cm. |
| 63 | 14 | 51.4 | +19 | 06 | ξ Boo | 4.7, 7.0 | 6.4 | 313 | Σ1888 | Fine contrast. Easy. |
| 64 | 15 | 03.8 | +47 | 39 | 44 Boo | 5.3, 6.2 | 2.2 | 56 | Σ1909 | Period = 246 years. |
| 65 | 15 | 05.1 | −47 | 03 | π Lup | 4.6, 4.7 | 1.7 | 64 | HJ 4728 | Widening. |
| 66 | 15 | 18.5 | −47 | 53 | μ Lup AB | 5.1, 5.2 | 0.8 | 120 | HJ 4753 | AB closing. Underobserved. |
| | | | | | μ Lup AC | 4.4, 7.2 | 24.0 | 129 | Δ180 | AC almost fixed. |
| 67 | 15 | 23.4 | −59 | 19 | γ Cir | 5.1, 5.5 | 0.8 | 352 | HJ 4757 | Closing. Needs 20 cm. Long-period binary. |
| 68 | 15 | 32.0 | +32 | 17 | η CrB | 5.6, 5.9 | 0.5 | 111 | Σ1937 | Both yellow. P = 41 yrs. Closing. |
| 69 | 15 | 34.8 | +10 | 33 | δ Ser | 4.2, 5.2 | 4.3 | 176 | Σ1954 | Long-period binary. |
| 70 | 15 | 35.1 | −41 | 10 | γ Lup | 3.5, 3.6 | 0.8 | 277 | HJ 4786 | Binary. Period = 190 years. Needs 20 cm. |
| 71 | 15 | 56.9 | −33 | 58 | ξ Lup | 5.3, 5.8 | 10.2 | 49 | PZ 4 | Fixed. |
| 72 | 16 | 14.7 | +33 | 52 | σ CrB | 5.6, 6.6 | 7.1 | 237 | Σ2032 | Long-period binary. Both white. |
| 73 | 16 | 29.4 | −26 | 26 | α Sco | 1.2, 5.4 | 2.6 | 274 | GNT 1 | Red, green. Difficult from mid northern latitudes. |
| 74 | 16 | 30.9 | +01 | 59 | λ Oph | 4.2, 5.2 | 1.5 | 34 | Σ2055 | P = 129 years. Fairly difficult in small apertures. |

| No. | RA | | Declin-ation | | Star | Magni-tudes | Separa-tion | PA | Cata-logue | Comments |
|---|---|---|---|---|---|---|---|---|---|---|
| | h | m | ° | ′ | | | arcsec | ° | | |
| 75 | 16 | 41.3 | +31 | 36 | ζ Her | 2.9, 5.5 | 0.9 | 225 | Σ2084 | Period 34 years. Now widening. Needs 30 cm. |
| 76 | 17 | 05.3 | +54 | 28 | μ Dra | 5.7, 5.7 | 2.3 | 12 | Σ2130 | Period 672 years. |
| 77 | 17 | 14.6 | +14 | 24 | α Her | var, 5.4 | 4.6 | 104 | Σ2140 | Red, green. Long-period binary. |
| 78 | 17 | 15.3 | −26 | 35 | 36 Oph | 5.1, 5.1 | 4.9 | 326 | SHJ 243 | Period = 471 years. |
| 79 | 17 | 23.7 | +37 | 08 | ρ Her | 4.6, 5.6 | 4.1 | 318 | Σ2161 | Slowly widening. |
| 80 | 18 | 01.5 | +21 | 36 | 95 Her | 5.0, 5.1 | 6.4 | 257 | Σ2264 | Colours thought variable in C19. |
| 81 | 18 | 05.5 | +02 | 30 | 70 Oph | 4.2, 6.0 | 4.9 | 138 | Σ2272 | Opening. Easy in 7.5 cm. |
| 82 | 18 | 06.8 | −43 | 25 | h5014 CrA | 5.7, 5.7 | 1.7 | 4 | – | Period = 450 years. Needs 10 cm. |
| 83 | 18 | 35.9 | +16 | 58 | OΣ358 Her | 6.8, 7.0 | 1.6 | 152 | – | Period = 380 years. |
| 84 | 18 | 44.3 | +39 | 40 | ε¹ Lyr | 5.0, 6.1 | 2.5 | 349 | Σ2382 | Quadruple system with ε². Both pairs |
| 85 | 18 | 44.3 | +39 | 40 | ε² Lyr | 5.2, 5.5 | 2.3 | 80 | Σ2383 | visible in 7.5 cm. |
| 86 | 18 | 56.2 | +04 | 12 | θ Ser | 4.5, 5.4 | 22.4 | 104 | Σ2417 | Fixed. Very easy. |
| 87 | 19 | 06.4 | −37 | 04 | γ CrA | 4.8, 5.1 | 1.3 | 37 | HJ 5084 | Beautiful pair. Period = 122 years. |
| 88 | 19 | 30.7 | +27 | 58 | β Cyg AB | 3.1, 5.1 | 34.3 | 54 | Σ I 43 | Glorious. Yellow, blue-greenish. |
| | | | | | β Cyg Aa | 3.1, 5.2 | 0.3 | 110 | MCA 55 | Aa. Period = 97 years. Closing. |
| 89 | 19 | 45.0 | +45 | 08 | δ Cyg | 2.9, 6.3 | 2.6 | 222 | Σ2579 | Slowly widening. Period = 780 years. |
| 90 | 19 | 48.2 | +70 | 16 | ε Dra | 3.8, 7.4 | 3.2 | 17 | Σ2603 | Slow binary. |

| No. | RA | | Declin-ation | | Star | Magni-tudes | Separa-tion | PA | Cata-logue | Comments |
|-----|----|----|----|----|------|------|------|----|----|----------|
| | h | m | ° | ′ | | | arcsec | ° | | |
| 91 | 20 | 46.7 | +16 | 07 | γ Del | 4.5, 5.5 | 9.2 | 266 | Σ2727 | Easy. Yellowish. Long-period binary. |
| 92 | 20 | 47.4 | +36 | 29 | λ Cyg | 4.8, 6.1 | 0.9 | 10 | OΣ413 | Difficult binary in small apertures. |
| 93 | 20 | 59.1 | +04 | 18 | ε Equ AB | 6.0, 6.3 | 0.7 | 284 | Σ2737 | Fine triple. AB is closing. |
| | | | | | ε Equ AC | 6.0, 7.1 | 10.3 | 66 | | |
| 94 | 21 | 06.9 | +38 | 45 | 61 Cyg | 5.2, 6.0 | 30.9 | 151 | Σ2758 | Nearby binary. Both orange. Period = 722 years. |
| 95 | 21 | 19.9 | −53 | 27 | θ Ind | 4.5, 7.0 | 6.8 | 271 | HJ 5258 | Pale yellow and reddish. Long-period binary. |
| 96 | 21 | 44.1 | +28 | 45 | μ Cyg | 4.8, 6.1 | 1.8 | 312 | Σ2822 | Period = 713 years. |
| 97 | 22 | 03.8 | +64 | 37 | ξ Cep | 4.4, 6.5 | 8.1 | 275 | Σ2863 | White and blue. Long-period binary. |
| 98 | 22 | 26.6 | −16 | 45 | 53 Aqr | 6.4, 6.6 | 1.5 | 20 | SHJ 345 | Long-period binary, approaching periastron. |
| 99 | 22 | 28.8 | −00 | 01 | ζ Aqr | 4.3, 4.5 | 2.1 | 178 | Σ2909 | Slowly widening. |
| 100 | 23 | 59.4 | +33 | 43 | Σ3050 And | 6.6, 6.6 | 2.1 | 332 | – | Period = 350 years. |

# Some Interesting Nebulae, Clusters and Galaxies

| Object | RA | | Declina-tion | | Remarks |
|---|---|---|---|---|---|
| | h | m | ° | ′ | |
| M31 Andromedae | 00 | 40.7 | +41 | 05 | Andromeda Galaxy, visible to naked eye. |
| H VIII 78 Cassiopeiae | 00 | 41.3 | +61 | 36 | Fine cluster, between Gamma and Kappa Cassiopeiae. |
| M33 Trianguli | 01 | 31.8 | +30 | 28 | Spiral. Difficult with small apertures. |
| H VI 33–4 Persei, C14 | 02 | 18.3 | +56 | 59 | Double cluster; Sword-handle. |
| Δ142 Doradus | 05 | 39.1 | −69 | 09 | Looped nebula round 30 Doradus. Naked eye. In Large Magellanic Cloud. |
| M1 Tauri | 05 | 32.3 | +22 | 00 | Crab Nebula, near Zeta Tauri. |
| M42 Orionis | 05 | 33.4 | −05 | 24 | Orion Nebula. Contains the famous Trapezium, Theta Orionis. |
| M35 Geminorum | 06 | 06.5 | +24 | 21 | Open cluster near Eta Geminorum. |
| H VII 2 Monocerotis, C50 | 06 | 30.7 | +04 | 53 | Open cluster, just visible to naked eye. |
| M41 Canis Majoris | 06 | 45.5 | −20 | 42 | Open cluster, just visible to naked eye. |
| M47 Puppis | 07 | 34.3 | −14 | 22 | Mag. 5.2. Loose cluster. |
| H IV 64 Puppis | 07 | 39.6 | −18 | 05 | Bright planetary in rich neighbourhood. |
| M46 Puppis | 07 | 39.5 | −14 | 42 | Open cluster. |
| M44 Cancri | 08 | 38 | +20 | 07 | Praesepe. Open cluster near Delta Cancri. Visible to naked eye. |
| M97 Ursae Majoris | 11 | 12.6 | +55 | 13 | Owl Nebula, diameter 3′. Planetary. |
| Kappa Crucis, C94 | 12 | 50.7 | −60 | 05 | 'Jewel Box'; open cluster, with stars of contrasting colours. |
| M3 Can. Ven. | 13 | 40.6 | +28 | 34 | Bright globular. |
| Omega Centauri, C80 | 13 | 23.7 | −47 | 03 | Finest of all globulars. Easy with naked eye. |
| M80 Scorpii | 16 | 14.9 | −22 | 53 | Globular, between Antares and Beta Scorpii. |
| M4 Scorpii | 16 | 21.5 | −26 | 26 | Open cluster close to Antares. |

| Object | RA | | Declina-tion | | Remarks |
|---|---|---|---|---|---|
| | h | m | ° | ′ | |
| M13 Herculis | 16 | 40 | +36 | 31 | Globular. Just visible to naked eye. |
| M92 Herculis | 16 | 16.1 | +43 | 11 | Globular. Between Iota and Eta Herculis. |
| M6 Scorpii | 17 | 36.8 | −32 | 11 | Open cluster; naked eye. |
| M7 Scorpii | 17 | 50.6 | −34 | 48 | Very bright open cluster; naked eye. |
| M23 Sagittarii | 17 | 54.8 | −19 | 01 | Open cluster nearly 50′ in diameter. |
| H IV 37 Draconis, C6 | 17 | 58.6 | +66 | 38 | Bright planetary. |
| M8 Sagittarii | 18 | 01.4 | −24 | 23 | Lagoon Nebula. Gaseous. Just visible with naked eye. |
| NGC 6572 Ophiuchi | 18 | 10.9 | +06 | 50 | Bright planetary, between Beta Ophiuchi and Zeta Aquilae. |
| M17 Sagittarii | 18 | 18.8 | −16 | 12 | Omega Nebula. Gaseous. Large and bright. |
| M11 Scuti | 18 | 49.0 | −06 | 19 | Wild Duck. Bright open cluster. |
| M57 Lyrae | 18 | 52.6 | +32 | 59 | Ring Nebula. Brightest of planetaries. |
| M27 Vulpeculae | 19 | 58.1 | +22 | 37 | Dumb-bell Nebula, near Gamma Sagittae. |
| H IV 1 Aquarii, C55 | 21 | 02.1 | −11 | 31 | Bright planetary, near Nu Aquarii. |
| M15 Pegasi | 21 | 28.3 | +12 | 01 | Bright globular, near Epsilon Pegasi. |
| M39 Cygni | 21 | 31.0 | +48 | 17 | Open cluster between Deneb and Alpha Lacertae. Well seen with low powers. |

(M = Messier number; NGC = New General Catalogue number; C = Caldwell number.)

# Our Contributors

**Dr Peter Cattermole** was formerly a lecturer in planetary geology and volcanology at the University of Sheffield and a Principal Investigator with NASA's Planetary Geology Program. He is now a freelance writer and lecturer, and spends much of his time leading specialist tour groups to all parts of the world for a London-based company.

**Dr Paul Murdin** was formerly Head of Astronomy at the Particle Physics and Astronomy Research Council (PPARC) and Director of Science at the British National Space Centre. He now works at the Institute of Astronomy in Cambridge. Paul is, of course, one of our most regular contributors.

**Dr Joe McCall** has worked as a professional geologist in Africa, Australia, Iran, Canada and the UK and has long embraced applications of geology to meteoritics and planetology. He curated the collection of meteorites at the Western Australian Museum for ten years while a Reader at the University in Perth. He is at present editing a history of meteoritics.

**Bob Argyle** has been Director of the Webb Society Double Star Section since 1970 and has recently edited a book on visual double star observing. By day he looks after the data archive for the UK La Palma Telescopes at the Institute of Astronomy, Cambridge.

**Professor Garry E. Hunt** has been a leading international space scientist for many decades with Viking (Mars), Voyager, where he was the only UK scientist involved with the mission, and an adviser to NASA, ESA. He has held senior appointments at Imperial College, UCL and JPL/CalTech, as well as many visiting Professorships in the US, Canada and Australia. He is now Managing Partner of Elbury Enterprises. Garry continues to be a regular contributor to BBC TV and *The Sky at Night* as well as the *Yearbook of Astronomy*.

**Professor Chris Kitchin** was formerly Director of the University of Hertfordshire Observatory. He is an astrophysicist with a great eagerness in encouraging a popular interest in astronomy. He is the author of several books, and appears regularly on television.

**Michael Maunder** describes himself as a chemist by profession, but an astronomer by inclination. He started chasing eclipses in 1973 when he joined Patrick Moore on the Monte Umbe for the first of the large expeditions.

**Professor Fred Watson** is Astronomer-in-Charge of the Anglo-Australian Observatory at Coonabarabran, New South Wales. He is an adjunct Professor in the School of Physical and Chemical Sciences of the Queensland University of Technology, and an honorary Associate Professor of Astronomy in the University of Southern Queensland. He is a member of the international RAVE project team, and a regular contributor to the *Yearbook of Astronomy*.

**Dr David M. Harland** gained his BSc in astronomy in 1977 and a doctorate in computational science. Subsequently, he has taught computer science, worked in industry and managed academic research. In 1995 he 'retired' and has since published many books on space themes.

**Dr Allan Chapman**, of Wadham College, Oxford, is probably Britain's leading authority on the history of astronomy. He has published many research papers and several books, as well as numerous popular accounts. He is a frequent and welcome contributor to the *Yearbook*.

# Astronomical Societies in the British Isles

**British Astronomical Association**
*Assistant Secretary:* Burlington House, Piccadilly, London W1V 9AG.
*Meetings:* Lecture Hall of Scientific Societies, Civil Service Commission Building, 23 Savile Row, London W1. Last Wednesday each month (Oct.–June), 5 p.m. and some Saturday afternoons.

**Association for Astronomy Education**
*Secretary:* Teresa Grafton, The Association for Astronomy Education, c/o The Royal Astronomical Society, Burlington House, Piccadilly, London W1V 0NL.

**Astronomical Society of Edinburgh**
*Secretary:* Graham Rule, 105/19 Causewayside, Edinburgh EH9 1QG.
*Web site:* www.roe.ac.uk/asewww/; *Email:* asewww@roe.ac.uk
*Meetings:* City Observatory, Calton Hill, Edinburgh. 1st Friday each month, 8 p.m.

**Astronomical Society of Glasgow**
*Secretary:* Mr David Degan, 5 Hillside Avenue, Alexandrdia, Dunbartonshire, G83 0BB.
*Web site:* www.astronomicalsocietyofglasgow.org.uk
*Meetings:* Royal College, University of Strathclyde, Montrose Street, Glasgow. 3rd Thursday each month, Sept.–Apr., 7.30 p.m.

**Astronomical Society of Haringey**
*Secretary:* Jerry Workman, 91 Greenslade Road, Barking, Essex, IG11 9XF.
*Meetings:* Palm Court, Alexandra Palace, 3rd Wednesday each month, 8 p.m.

**Astronomy Ireland**
*Secretary:* Tony Ryan, PO Box 2888, Dublin 1, Eire.
*Web site:* www.astronomy.ie; *Email:* info@astronomy.ie
*Meetings:* 2nd Monday of each month. Telescope meetings every clear Saturday.

**Federation of Astronomical Societies**
*Secretary:* Clive Down, 10 Glan-y-Llyn, North Cornelly, Bridgend, County Borough, CF33 4EF.
*Email:* clivedown@btinternet.com

**Junior Astronomical Society of Ireland**
*Secretary:* K. Nolan, 5 St Patrick's Crescent, Rathcoole, Co. Dublin.
*Meetings:* The Royal Dublin Society, Ballsbridge, Dublin 4. Monthly.

**Society for Popular Astronomy**
*Secretary:* Guy Fennimore, 36 Fairway, Keyworth, Nottingham, NG12 5DU.
*Web site:* www.popastro.com; *Email:* SPAstronomy@aol.com
*Meetings:* Last Saturday in Jan., Apr., July, Oct., 2.30 p.m. in London.

**Webb Society**
*Secretary:* M. B. Swan, Carrowreagh, Kilshanny, Kilfenora, Co. Clare, Eire.

**Aberdeen and District Astronomical Society**
*Secretary:* Ian C. Giddings, 95 Brentfield Circle, Ellon, Aberdeenshire AB41 9DB.
*Meetings:* Robert Gordon's Institute of Technology, St Andrew's Street, Aberdeen.
Fridays, 7.30 p.m.

**Abingdon Astronomical Society** (was **Fitzharry's Astronomical Society**)
*Secretary:* Chris Holt, 9 Rutherford Close, Abingdon, Oxon OX14 2AT.
*Web site:* www.abingdonastro.org.uk; *Email:* info@abingdonastro.co.uk
*Meetings:* All Saints' Methodist Church Hall, Dorchester Crescent, Abingdon, Oxon.
2nd Monday Sept.–June, 8 p.m. and additional beginners' meetings and observing
evenings as advertised.

**Altrincham and District Astronomical Society**
*Secretary:* Derek McComiskey, 33 Tottenham Drive, Manchester M23 9WH.
*Meetings:* Timperley Village Club. 1st Friday Sept.–June, 8 p.m.

**Andover Astronomical Society**
*Secretary:* Mrs S. Fisher, Staddlestones, Aughton, Kingston, Marlborough, Wiltshire,
SN8 3SA.
*Meetings:* Grately Village Hall. 3rd Thursday each month, 7.30 p.m.

**Astra Astronomy Section**
*Secretary:* c/o Duncan Lunan, Flat 65, Dalraida House, 56 Blythswood Court,
Anderston, Glasgow G2 7PE.
*Meetings:* Airdrie Arts Centre, Anderson Street, Airdrie. Weekly.

**Astrodome Mobile School Planetarium**
*Contact:* Peter J. Golding, 53 City Way, Rochester, Kent ME1 2AX.
*Web site:* www.astrodome.clara.co.uk; *Email:* astrodome@clara.co.uk

**Aylesbury Astronomical Society**
*Secretary:* Alan Smith, 182 Marley Fields, Leighton Buzzard, Beds, LU7 8WN.
*Meetings:* 1st Monday in month at 8 p.m., venue in Aylesbury area. Details from
Secretary.

**Bassetlaw Astronomical Society**
*Secretary:* Andrew Patton, 58 Holding, Worksop, Notts S81 0TD.
*Meetings:* Rhodesia Village Hall, Rhodesia, Worksop, Notts. 2nd and 4th Tuesdays of
month at 7.45 p.m.

**Batley & Spenborough Astronomical Society**
*Secretary:* Robert Morton, 22 Links Avenue, Cleckheaton, West Yorks BD19 4EG.
*Meetings:* Milner K. Ford Observatory, Wilton Park, Batley. Every Thursday, 8 p.m.

**Bedford Astronomical Society**
*Secretary:* Mrs L. Harrington, 24 Swallowfield, Wyboston, Bedfordshire, MK44 3AE.
*Web site:* www.observer1.freeserve.co.uk/bashome.html
*Meetings:* Bedford School, Burnaby Rd, Bedford. Last Wednesday each month.

**Bingham & Brooks Space Organization**
*Secretary:* N. Bingham, 15 Hickmore's Lane, Lindfield, W. Sussex.

**Birmingham Astronomical Society**
*Contact:* P. Bolas, 4 Moat Bank, Bretby, Burton on Trent DE15 0QJ.
*Web site:* www.birmingham-astronomical.co.uk; *Email:* pbolas@aol.com
*Meetings:* Room 146, Aston University. Last Tuesday of month. Sept.–June (except
Dec., moved to 1st week in Jan.).

**Blackburn Leisure Astronomy Section**
*Secretary:* Mr H. Murphy, 20 Princess Way, Beverley, East Yorkshire, HU17 8PD.
*Meetings:* Blackburn Leisure Welfare. Mondays, 8 p.m.

**Blackpool & District Astronomical Society**
*Secretary:* Terry Devon, 30 Victory Road, Blackpool, Lancashire, FY1 3JT.
*Acting Secretary:* Tony Evanson, 25 Aintree Road, Thornton, Lancashire, FY5 5HW.
*Web site:* www.geocities.com/bad_astro/index.html; *Email:* bad_astro@yahoo.co.uk
*Meetings:* St Kentigens Social Centre, Blackpool. 1st Wednesday of the month, 8 p.m.

**Bolton Astronomical Society**
*Secretary:* Peter Miskiw, 9 Hedley Street, Bolton, Lancashire, BL1 3LE.
*Meetings:* Ladybridge Community Centre, Bolton. 1st and 3rd Tuesdays Sept.–May,
7.30 p.m.

**Border Astronomy Society**
*Secretary:* David Pettitt, 14 Sharp Grove, Carlisle, Cumbria, CA2 5QR.
*Web site:* www.members.aol.com/P3pub/page8.html
*Email:* davidpettitt@supanet.com
*Meetings:* The Observatory, Trinity School, Carlisle. Alternate Thursdays, 7.30 p.m.,
Sept.–May.

**Boston Astronomers**
*Secretary:* Mrs Lorraine Money, 18 College Park, Horncastle, Lincolnshire, LN9 6RE.
*Meetings:* Blackfriars Arts Centre, Boston. 2nd Monday each month, 7.30 p.m.

**Bradford Astronomical Society**
*Contact:* Mrs J. Hilary Knaggs, 6 Meadow View, Wyke, Bradford, BD12 9LA.
*Web site:* www.bradford-astro.freeserve.co.uk/index.htm
*Meetings:* Eccleshill Library, Bradford. Alternate Mondays, 7.30 p.m.

**Braintree, Halstead & District Astronomical Society**
*Secretary:* Mr J. R. Green, 70 Dorothy Sayers Drive, Witham, Essex, CM8 2LU.
*Meetings:* BT Social Club Hall, Witham Telephone Exchange. 3rd Thursday each
month, 8 p.m.

**Breckland Astronomical Society** (was **Great Ellingham and District Astronomy Club**)
*Contact:* Martin Wolton, Willowbeck House, Pulham St Mary, Norfolk, IP21 4QS.
*Meetings:* Great Ellingham Recreation Centre, Watton Road (B1077), Great
Ellingham, 2nd Friday each month, 7.15 p.m.

**Bridgend Astronomical Society**
*Secretary:* Clive Down, 10 Glan-y-Llyn, Broadlands, North Cornelly, Bridgend
County, CF33 4EF.
*Email:* clivedown@btinternet.com
*Meetings:* Bridgend Bowls Centre, Bridgend. 2nd Friday, monthly, 7.30 p.m.

**Bridgwater Astronomical Society**
*Secretary:* Mr G. MacKenzie, Watergore Cottage, Watergore, South Petherton,
Somerset, TA13 5JQ.
*Web site:* www.ourworld.compuserve.com/hompages/dbown/Bwastro.htm
*Meetings:* Room D10, Bridgwater College, Bath Road Centre, Bridgwater. 2nd
Wednesday each month, Sept.–June.

**Bridport Astronomical Society**
*Secretary:* Mr G. J. Lodder, 3 The Green, Walditch, Bridport, Dorset, DT6 4LB.
*Meetings:* Walditch Village Hall, Bridport. 1st Sunday each month, 7.30 p.m.

**Brighton Astronomical and Scientific Society**
*Secretary:* Ms T. Fearn, 38 Woodlands Close, Peacehaven, East Sussex, BN10 7SF.
*Meetings:* St Johns Church Hall, Hove. 1st Tuesday each month, 7.30 p.m.

**Bristol Astronomical Society**
*Secretary:* Dr John Pickard, 'Fielding', Easter Compton, Bristol, BS35 5SJ.
*Meetings:* Frank Lecture Theatre, University of Bristol Physics Dept., alternate Fridays in term time, and Westbury Park Methodist Church Rooms, North View, other Fridays.

**Callington Community Astronomy Group**
*Secretary:* Beccy Watson. Tel: 07732 945671
*Email:* Beccyboo@kimwatson99.fsnet.co.uk
*Website:* www.callington-astro.org.uk
*Meetings:* Callington Space Centre, Callington Community College, Launceston Road, Callington, Cornwall, PL17 7DR. 1st and 3rd Saturday of each month, 7.30 p.m., Sept.–July.

**Cambridge Astronomical Society**
*Secretary:* Brian Lister, 80 Ramsden Square, Cambridge CB4 2BL.
*Meetings:* Institute of Astronomy, Madingley Road. 3rd Friday each month.

**Cardiff Astronomical Society**
*Secretary:* D. W. S. Powell, 1 Tal-y-Bont Road, Ely, Cardiff CF5 5EU.
*Meetings:* Dept. of Physics and Astronomy, University of Wales, Newport Road, Cardiff. Alternate Thursdays, 8 p.m.

**Castle Point Astronomy Club**
*Secretary:* Andrew Turner, 3 Canewdon Hall Close, Canewdon, Rochford, Essex SS4 3PY.
*Meetings:* St Michael's Church Hall, Daws Heath. Wednesdays, 8 p.m.

**Chelmsford Astronomers**
*Secretary:* Brendan Clark, 5 Borda Close, Chelmsford, Essex.
*Meetings:* Once a month.

**Chester Astronomical Society**
*Secretary:* Mrs S. Brooks, 39 Halton Road, Great Sutton, South Wirral, LL66 2UF.
*Meetings:* All Saints Parish Church, Chester. Last Wednesday each month except Aug. and Dec., 7.30 p.m.

**Chester Society of Natural Science, Literature and Art**
*Secretary:* Paul Braid, 'White Wing', 38 Bryn Avenue, Old Colwyn, Colwyn Bay LL29 8AH.
*Email:* p.braid@virgin.net
*Meetings:* Once a month.

**Chesterfield Astronomical Society**
*President:* Mr D. Blackburn, 71 Middlecroft Road, Stavely, Chesterfield, Derbyshire, S41 3XG. Tel: 07909 570754.
*Website:* www.chesterfield-as.org.uk
*Meetings:* Barnet Observatory, Newbold, each Friday.

**Clacton & District Astronomical Society**
*Secretary:* C. L. Haskell, 105 London Road, Clacton-on-Sea, Essex.

**Cleethorpes & District Astronomical Society**
*Secretary:* C. Illingworth, 38 Shaw Drive, Grimsby, S. Humberside.
*Meetings:* Beacon Hill Observatory, Cleethorpes. 1st Wednesday each month.

**Cleveland & Darlington Astronomical Society**
*Contact:* Dr. John McCue, 40 Bradbury Rd., Stockton-on-Tees, Cleveland TS20 1LE.
*Meetings:* Grindon Parish Hall, Thorpe Thewles, near Stockton-on-Tees. 2nd Friday, monthly.

**Cork Astronomy Club**
*Secretary:* Charles Coughlan, 12 Forest Ridge Crescent, Wilton, Cork, Eire.
*Meetings:* 1st Monday, Sept.–May (except bank holidays).
**Cornwall Astronomical Society**
*Secretary:* J. M. Harvey, 1 Tregunna Close, Porthleven, Cornwall TR13 9LW.
*Meetings:* Godolphin Club, Wendron Street, Helston, Cornwall. 2nd and 4th
Thursday of each month, 7.30 for 8 p.m.
**Cotswold Astronomical Society**
*Secretary:* Rod Salisbury, Grove House, Christchurch Road, Cheltenham, Glos
GL50 2PN.
*Web site:* www.members.nbci.com/CotswoldAS
*Meetings:* Shurdington Church Hall, School Lane, Shurdington, Cheltenham. 2nd
Saturday each month, 8 p.m.
**Coventry & Warwickshire Astronomical Society**
*Secretary:* Steve Payne, 68 Stonebury Avenue, Eastern Green, Coventry CV5 7FW.
*Web site:* www.cawas.freeserve.co.uk; *Email:* sjp2000@thefarside57.freeserve.co.uk
*Meetings:* The Earlsdon Church Hall, Albany Road, Earlsdon, Coventry. 2nd Friday,
monthly, Sept.–June.
**Crawley Astronomical Society**
*Secretary:* Ron Gamer, 1 Pevensey Close, Pound Hill, Crawley, West Sussex
RH10 7BL.
*Meetings:* Ifield Community Centre, Ifield Road, Crawley. 3rd Friday each month,
7.30 p.m.
**Crayford Manor House Astronomical Society**
*Secretary:* Roger Pickard, 28 Appletons, Hadlow, Kent TM1 0DT.
*Meetings:* Manor House Centre, Crayford. Monthly during term time.
**Crewkerne and District Astronomical Society (CADAS)**
*Chairman:* Kevin Dodgson, 46 Hermitage Street, Crewkerne, Somerset, TA18 8ET.
*Email:* crewastra@aol.com
**Croydon Astronomical Society**
*Secretary:* John Murrell, 17 Dalmeny Road, Carshalton, Surrey.
*Meetings:* Lecture Theatre, Royal Russell School, Combe Lane, South Croydon.
Alternate Fridays, 7.45 p.m.
**Derby & District Astronomical Society**
*Secretary:* Ian Bennett, Freers Cottage, Sutton Lane, Etwall.
*Web site:* www.derby-astro-soc.fsnet/index.html
*Email:* bennett.lovatt@btinternet.com
*Meetings:* Friends Meeting House, Derby. 1st Friday each month, 7.30 p.m.
**Doncaster Astronomical Society**
*Secretary:* A. Anson, 15 Cusworth House, St James Street, Doncaster, DN1 3AY
*Web site:* www.donastro.freeserve.co.uk
*Email:* space@donastro.freeserve.co.uk
*Meetings:* St George's Church House, St George's Church, Church Way, Doncaster.
2nd and 4th Thursday of each month, commencing at 7.30 p.m.
**Dumfries Astronomical Society**
*Secretary:* Mr J. Sweeney, 3 Lakeview, Powfoot, Annan, DG13 5PG.
*Meetings:* Gracefield Arts Centre, Edinburgh Road, Dumfries. 3rd Tuesday Aug.–
May, 7.30 p.m.

**Dundee Astronomical Society**
> *Secretary:* G. Young, 37 Polepark Road, Dundee, Tayside, DD1 5QT.
> *Meetings:* Mills Observatory, Balgay Park, Dundee. 1st Friday each month, 7.30 p.m.
> Sept.–Apr.

**Easington and District Astronomical Society**
> *Secretary:* T. Bradley, 52 Jameson Road, Hartlepool, Co. Durham.
> *Meetings:* Easington Comprehensive School, Easington Colliery. Every 3rd Thursday
> throughout the year, 7.30 p.m.

**Eastbourne Astronomical Society**
> *Secretary:* Peter Gill, 18 Selwyn House, Selwyn Road, Eastbourne, East Sussex
> BN21 2LF.
> *Meetings:* Willingdon Memorial Hall, Church Street, Willingdon. One Saturday per
> month, Sept.–July, 7.30 p.m.

**East Riding Astronomers**
> *Secretary:* Tony Scaife, 15 Beech Road, Elloughton, Brough, North Humberside,
> HU15 1JX.
> *Meetings:* As arranged.

**East Sussex Astronomical Society**
> *Secretary:* Marcus Croft, 12 St Marys Cottages, Ninfield Road, Bexhill on Sea, East
> Sussex.
> *Web site:* www.esas.org.uk
> *Meetings:* St Marys School, Wrestwood Road, Bexhill. 1st Thursday of each month,
> 8 p.m.

**Edinburgh University Astronomical Society**
> *Secretary:* c/o Dept. of Astronomy, Royal Observatory, Blackford Hill, Edinburgh.

**Ewell Astronomical Society**
> *Secretary:* Richard Gledhill, 80 Abinger Avenue, Cheam SM2 7LW.
> *Web site:* www.ewell-as.co.uk
> *Meetings:* St Mary's Church Hall, London Road, Ewell. 2nd Friday of each month
> except August, 7.45 p.m.

**Exeter Astronomical Society**
> *Secretary:* Tim Sedgwick, Old Dower House, Half Moon, Newton St Cyres, Exeter,
> Devon, EX5 5AE.
> *Meetings:* The Meeting Room, Wynards, Magdalen Street, Exeter. 1st Thursday of
> month.

**Farnham Astronomical Society**
> *Secretary:* Laurence Anslow, 'Asterion', 18 Wellington Lane, Farnham, Surrey,
> GU9 9BA.
> *Meetings:* Central Club, South Street, Farnham. 2nd Thursday each month, 8 p.m.

**Foredown Tower Astronomy Group**
> *Secretary:* M. Feist, Foredown Tower Camera Obscura, Foredown Road, Portslade,
> East Sussex BN41 2EW.
> *Meetings:* At the above address, 3rd Tuesday each month. 7 p.m. (winter), 8 p.m.
> (summer).

**Fylde Astronomical Society**
> *Secretary:* 28 Belvedere Road, Thornton, Lancs.
> *Meetings:* Stanley Hall, Rossendale Avenue South. 1st Wednesday each month.

**Greenock Astronomical Society**
*Secretary:* Carl Hempsey, 49 Brisbane Street, Greenock.
*Meetings:* Greenock Arts Guild, 3 Campbell Street, Greenock.

**Grimsby Astronomical Society**
*Secretary:* R. Williams, 14 Richmond Close, Grimsby, South Humberside.
*Meetings:* Secretary's home. 2nd Thursday each month, 7.30 p.m.

**Guernsey: La Société Guernesiasie Astronomy Section**
*Secretary:* Debby Quertier, Lamorna, Route Charles, St Peter Port, Guernsey GY1
1QS and Jessica Harris, Keanda, Les Sauvagees, St Sampsons, Guernsey GY2 4XT.
*Meetings:* Observatory, Rue du Lorier, St Peters. Tuesdays, 8 p.m.

**Guildford Astronomical Society**
*Secretary:* A. Langmaid, 22 West Mount, The Mount, Guildford, Surrey, GU2 5HL.
*Meetings:* Guildford Institute, Ward Street, Guildford. 1st Thursday each month,
except Aug., 7.30 p.m.

**Gwynedd Astronomical Society**
*Secretary:* Mr Ernie Greenwood, 18 Twrcelyn Street, Llanerchymedd, Anglesey
LL74 8TL.
*Meetings:* Dept. of Electronic Engineering, Bangor University. 1st Thursday each
month except Aug., 7.30 p.m.

**The Hampshire Astronomical Group**
*Secretary:* Geoff Mann, 10 Marie Court, 348 London Road, Waterlooville, Hants
PO7 7SR.
*Web site:* www.hantsastro.demon.co.uk; *Email:* Geoff.Mann@hazleton97.fsnet.co.uk
*Meetings:* 2nd Friday, Clanfield Memorial Hall, all other Fridays Clanfield
Observatory.

**Hanney & District Astronomical Society**
*Secretary:* Bob Church, 47 Upthorpe Drive, Wantage, Oxfordshire, OX12 7DG.
*Meetings:* Last Thursday each month, 8 p.m.

**Harrogate Astronomical Society**
*Secretary:* Brian Bonser, 114 Main Street, Little Ouseburn, TO5 9TG.
*Meetings:* National Power HQ, Beckwith Knowle, Harrogate. Last Friday each
month.

**Hastings and Battle Astronomical Society**
*Secretary:* K. A. Woodcock, 24 Emmanuel Road, Hastings, East Sussex, TN34 3LB.
*Email:* keith@habas.freeserve.co.uk
*Meetings:* Herstmonceux Science Centre. 2nd Saturday of each month, 7.30 p.m.

**Havering Astronomical Society**
*Secretary:* Frances Ridgley, 133 Severn Drive, Upminster, Essex, RM14 1PP.
*Meetings:* Cranham Community Centre, Marlborough Gardens, Upminster, Essex.
3rd Wednesday each month (except July and Aug.), 7.30 p.m.

**Heart of England Astronomical Society**
*Secretary:* John Williams, 100 Stanway Road, Shirley, Solihull, B90 3JG.
*Web site:* www.members.aol.com/hoeas/home.html; *Email:* hoeas@aol.com
*Meetings:* Furnace End Village, over Whitacre, Warwickshire. Last Thursday each
month, except June, July & Aug., 8 p.m.

**Hebden Bridge Literary & Scientific Society, Astronomical Section**
*Secretary:* Peter Jackson, 44 Gilstead Lane, Bingley, West Yorkshire, BD16 3NP.
*Meetings:* Hebden Bridge Information Centre. Last Wednesday, Sept.–May.

**Herschel Astronomy Society**
> *Secretary:* Kevin Bishop, 106 Holmsdale, Crown Wood, Bracknell, Berkshire, RG12 3TB.
> *Meetings:* Eton College. 2nd Friday each month, 7.30 p.m.

**Highlands Astronomical Society**
> *Secretary:* Richard Green, 11 Drumossie Avenue, Culcabock, Inverness IV2 3SJ.
> *Meetings:* The Spectrum Centre, Inverness. 1st Tuesday each month, 7.30 p.m.

**Hinckley & District Astronomical Society**
> *Secretary:* Mr S. Albrighton, 4 Walnut Close, The Bridleways, Hartshill, Nuneaton, Warwickshire, CV10 0XH.
> *Meetings:* Burbage Common Visitors Centre, Hinckley. 1st Tuesday Sept.–May, 7.30 p.m.

**Horsham Astronomy Group** (was **Forest Astronomical Society**)
> *Secretary:* Dan White, 32 Burns Close, Horsham, West Sussex, RH12 5PF.
> *Email:* secretary@horshamastronomy.com
> *Meetings:* 1st Wednesday each month.

**Howards Astronomy Club**
> *Secretary:* H. Ilett, 22 St Georges Avenue, Warblington, Havant, Hants.
> *Meetings:* To be notified.

**Huddersfield Astronomical and Philosophical Society**
> *Secretary:* Lisa B. Jeffries, 58 Beaumont Street, Netherton, Huddersfield, West Yorkshire, HD4 7HE.
> *Email:* l.b.jeffries@hud.ac.uk
> *Meetings:* 4a Railway Street, Huddersfield. Every Wednesday and Friday, 7.30 p.m.

**Hull and East Riding Astronomical Society**
> *President:* Rob Overfield, 125 Marlborough Avenue, Princes Avenue, Hull, HU5 3JU.
> *Email:* rob.overfield@btinternet.com
> *Meetings:* University of Hull, Cottingham Road, Hull: The Wilberforce Building, Room SR110. 1st Monday each month, Sept.–Apr., 7.30–10.00 p.m.

**Ilkeston & District Astronomical Society**
> *Secretary:* Mark Thomas, 2 Elm Avenue, Sandiacre, Nottingham NG10 5EJ.
> *Meetings:* The Function Room, Erewash Museum, Anchor Row, Ilkeston. 2nd Tuesday monthly, 7.30 p.m.

**Ipswich, Orwell Astronomical Society**
> *Secretary:* R. Gooding, 168 Ashcroft Road, Ipswich.
> *Meetings:* Orwell Park Observatory, Nacton, Ipswich. Wednesdays, 8 p.m.

**Irish Astronomical Association**
> *Secretary:* Terry Moseley (President), 31 Sunderland Road, Belfast BT6 9LY, N. Ireland.
> *Email:* terrymosel@aol.com
> *Meetings:* Ashby Building, Stranmillis Road, Belfast. Alternate Wednesdays, 7.30 p.m.

**Irish Astronomical Society**
> *Secretary:* James O'Connor, PO Box 2547, Dublin 15, Ireland.
> *Meetings:* Ely House, 8 Ely Place, Dublin 2. 1st and 3rd Monday each month.

**Isle of Man Astronomical Society**
> *Secretary:* James Martin, Ballaterson Farm, Peel, Isle of Man IM5 3AB.
> *Email:* ballaterson@manx.net
> *Meetings:* Isle of Man Observatory, Foxdale. 1st Thursday of each month, 8 p.m.

**Isle of Wight Astronomical Society**
  *Secretary:* J. W. Feakins, 1 Hilltop Cottages, High Street, Freshwater, Isle of Wight.
  *Meetings:* Unitarian Church Hall, Newport, Isle of Wight. Monthly.
**Keele Astronomical Society**
  *Secretary:* Natalie Webb, Department of Physics, University of Keele, Keele,
  Staffordshire, ST5 5BG.
  *Meetings:* As arranged during term time.
**Kettering and District Astronomical Society**
  *Asst. Secretary:* Steve Williams, 120 Brickhill Road, Wellingborough, Northants.
  *Meetings:* Quaker Meeting Hall, Northall Street, Kettering, Northants. 1st Tuesday
  each month, 7.45 p.m.
**King's Lynn Amateur Astronomical Association**
  *Secretary:* P. Twynman, 17 Poplar Avenue, RAF Marham, King's Lynn.
  *Meetings:* As arranged.
**Lancaster and Morecambe Astronomical Society**
  *Secretary:* Mrs E. Robinson, 4 Bedford Place, Lancaster, LA1 4EB.
  *Email:* ehelenerob@btinternet.com
  *Meetings:* Church of the Ascension, Torrisholme. 1st Wednesday each month, except
  July and Aug.
**Lancaster University Astronomical Society**
  *Secretary:* c/o Students Union, Alexandra Square, University of Lancaster.
  *Meetings:* As arranged.
**Laymans Astronomical Society**
  *Secretary:* John Evans, 10 Arkwright Walk, The Meadows, Nottingham.
  *Meetings:* The Popular, Bath Street, Ilkeston, Derbyshire. Monthly.
**Leeds Astronomical Society**
  *Secretary:* Mark A. Simpson, 37 Roper Avenue, Gledhow, Leeds, LS8 1LG.
  *Meetings:* Centenary House, North Street. 2nd Wednesday each month, 7.30 p.m.
**Leicester Astronomical Society**
  *Secretary:* Dr P. J. Scott, 21 Rembridge Close, Leicester LE3 9AP.
  *Meetings:* Judgemeadow Community College, Marydene Drive, Evington, Leicester.
  2nd and 4th Tuesdays each month, 7.30 p.m.
**Letchworth and District Astronomical Society**
  *Secretary:* Eric Hutton, 14 Folly Close, Hitchin, Herts.
  *Meetings:* As arranged.
**Lewes Amateur Astronomers**
  *Secretary:* Christa Sutton, 8 Tower Road, Lancing, West Sussex, BN15 9HT.
  *Meetings:* The Bakehouse Studio, Lewes. Last Wednesday each month.
**Limerick Astronomy Club**
  *Secretary:* Tony O'Hanlon, 26 Ballycannon Heights, Meelick, Co. Clare, Eire.
  *Meetings:* Limerick Senior College, Limerick, Ireland. Monthly (except June and
  Aug.), 8 p.m.
**Lincoln Astronomical Society**
  *Secretary:* David Swaey, 'Everglades', 13 Beaufort Close, Lincoln LN2 4SF.
  *Meetings:* The Lecture Hall, off Westcliffe Street, Lincoln. 1st Tuesday each month.
**Liverpool Astronomical Society**
  *Secretary:* Mr K. Clark, 31 Sandymount Drive, Wallasey, Merseyside L45 0LJ.
  *Meetings:* Lecture Theatre, Liverpool Museum. 3rd Friday each month, 7 p.m.

**Norman Lockyer Observatory Society**
*Secretary:* G. E. White, PO Box 9, Sidmouth EX10 0YQ.
*Web site:* www.ex.ac.uk/nlo/; *Email:* g.e.white@ex.ac.uk
*Meetings:* Norman Lockyer Observatory, Sidmouth. Fridays and 2nd Monday each month, 7.30 p.m.

**Loughton Astronomical Society**
*Secretary:* Charles Munton, 14a Manor Road, Wood Green, London N22 4YJ.
*Meetings:* 1st Theydon Bois Scout Hall, Loughton Lane, Theydon Bois. Weekly.

**Lowestoft and Great Yarmouth Regional Astronomers (LYRA) Society**
*Secretary:* Simon Briggs, 28 Sussex Road, Lowestoft, Suffolk.
*Meetings:* Community Wing, Kirkley High School, Kirkley Run, Lowestoft. 3rd Thursday each month, 7.30 p.m.

**Luton Astronomical Society**
*Secretary:* Mr G. Mitchell, Putteridge Bury, University of Luton, Hitchin Road, Luton.
*Web site:* www.lutonastrosoc.org.uk; *Email:* user998491@aol.com
*Meetings:* Putteridge Bury, Luton. Last Friday each month, 7.30 p.m.

**Lytham St Annes Astronomical Association**
*Secretary:* K. J. Porter, 141 Blackpool Road, Ansdell, Lytham St Annes, Lancs.
*Meetings:* College of Further Education, Clifton Drive South, Lytham St Annes. 2nd Wednesday monthly Oct.–June.

**Macclesfield Astronomical Society**
*Secretary:* Mr John H. Thomson, 27 Woodbourne Road, Sale, Chesire M33 3SY
*Web site:* www.g0-evp.demon.co.uk; *Email:* jhandlc@yahoo.com
*Meetings:* Jodrell Bank Science Centre, Goostrey, Cheshire. 1st Tuesday of every month, 7 p.m.

**Maidenhead Astronomical Society**
*Secretary:* Tim Haymes, Hill Rise, Knowl Hill Common, Knowl Hill, Reading RG10 9YD.
*Meetings:* Stubbings Church Hall, near Maidenhead. 1st Friday Sept.–June.

**Maidstone Astronomical Society**
*Secretary:* Stephen James, 4 The Cherry Orchard, Haddow, Tonbridge, Kent.
*Meetings:* Nettlestead Village Hall. 1st Tuesday in the month except July and Aug., 7.30 p.m.

**Manchester Astronomical Society**
*Secretary:* Mr Kevin J. Kilburn FRAS, Godlee Observatory, UMIST, Sackville Street, Manchester M60 1QD.
*Web site:* www.u-net.com/ph/mas/; *Email:* kkilburn@globalnet.co.uk
*Meetings:* At the Godlee Observatory. Thursdays, 7 p.m., except below.
Free Public Lectures: Renold Building UMIST, third Thursday Sept.–Mar., 7.30 p.m.

**Mansfield and Sutton Astronomical Society**
*Secretary:* Angus Wright, Sherwood Observatory, Coxmoor Road, Sutton-in-Ashfield, Nottinghamshire NG17 5LF.
*Meetings:* Sherwood Observatory, Coxmoor Road. Last Tuesday each month, 7.30 p.m.

**Mexborough and Swinton Astronomical Society**
*Secretary:* Mark R. Benton, 14 Sandalwood Rise, Swinton, Mexborough, South Yorkshire, S64 8PN.
*Web site:* www.msas.org.uk; *Email:* mark@masas.f9.co.uk
*Meetings:* Swinton WMC. Thursdays, 7.30 p.m.

**Mid-Kent Astronomical Society**
*Secretary:* Peter Bassett, 167 Shakespeare Road, Gillingham, Kent, ME7 5QB.
*Meetings:* Riverside Country Park, Lower Rainham Road, Gillingham. 2nd and last Fridays each month, 7.45 p.m.

**Milton Keynes Astronomical Society**
*Secretary:* Mike Leggett, 19 Matilda Gardens, Shenley Church End, Milton Keynes, MK5 6HT.
*Web site:* www.mkas.org.uk; *Email:* mike-pat-leggett@shenley9.fsnet.co.uk
*Meetings:* Rectory Cottage, Bletchley. Alternate Fridays.

**Moray Astronomical Society**
*Secretary:* Richard Pearce, 1 Forsyth Street, Hopeman, Elgin, Moray, Scotland.
*Meetings:* Village Hall Close, Co. Elgin.

**Newbury Amateur Astronomical Society**
*Secretary:* Miss Nicola Evans, 'Romaron', Bunces Lane, Burghfield Common, Reading RG7 3DG.
*Meetings:* United Reformed Church Hall, Cromwell Place, Newbury. 2nd Friday of month, Sept.–June.

**Newcastle-on-Tyne Astronomical Society**
*Secretary:* C. E. Willits, 24 Acomb Avenue, Seaton Delaval, Tyne and Wear.
*Meetings:* Zoology Lecture Theatre, Newcastle University. Monthly.

**North Aston Space & Astronomical Club**
*Secretary:* W. R. Chadburn, 14 Oakdale Road, North Aston, Sheffield.
*Meetings:* To be notified.

**Northamptonshire Natural History Society (Astronomy Section)**
*Secretary:* R. A. Marriott, 24 Thirlestane Road, Northampton NN4 8HD.
*Email:* ram@hamal.demon.co.uk
*Meetings:* Humfrey Rooms, Castilian Terrace, Northampton. 2nd and last Mondays, most months, 7.30 p.m.

**Northants Amateur Astronomers**
*Secretary:* Mervyn Lloyd, 76 Havelock Street, Kettering, Northamptonshire.
*Meetings:* 1st and 3rd Tuesdays each month, 7.30 p.m.

**North Devon Astronomical Society**
*Secretary:* P. G. Vickery, 12 Broad Park Crescent, Ilfracombe, Devon, EX34 8DX.
*Meetings:* Methodist Hall, Rhododendron Avenue, Sticklepath, Barnstaple. 1st Wednesday each month, 7.15 p.m.

**North Dorset Astronomical Society**
*Secretary:* J. E. M. Coward, The Pharmacy, Stalbridge, Dorset.
*Meetings:* Charterhay, Stourton, Caundle, Dorset. 2nd Wednesday each month.

**North Downs Astronomical Society**
*Secretary:* Martin Akers, 36 Timber Tops, Lordswood, Chatham, Kent, ME5 8XQ.
*Meetings:* Vigo Village Hall. 3rd Thursday each month. 7.30 p.m.

**North-East London Astronomical Society**
*Secretary:* Mr B. Beeston, 38 Abbey Road, Bush Hill Park, Enfield EN1 2QN.
*Meetings:* Wanstead House, The Green, Wanstead. 3rd Sunday each month (except Aug.), 3 p.m.

**North Gwent and District Astronomical Society**
*Secretary:* Jonathan Powell, 14 Lancaster Drive, Gilwern, nr Abergavenny, Monmouthshire, NP7 0AA.
*Meetings:* Gilwern Community Centre. 15th of each month, 7.30 p.m.

**North Staffordshire Astronomical Society**
*Secretary:* Duncan Richardson, Halmerend Hall Farm, Halmerend, Stoke-on-Trent, Staffordshire, ST7 8AW.
*Email:* dwr@enterprise.net
*Meetings:* 21st Hartstill Scout Group HQ, Mount Pleasant, Newcastle-under-Lyme ST5 1DR. 1st Tuesday each month (except July and Aug.), 7–9.30 p.m.

**North Western Association of Variable Star Observers**
*Secretary:* Jeremy Bullivant, 2 Beaminster Road, Heaton Mersey, Stockport, Cheshire.
*Meetings:* Four annually.

**Norwich Astronomical Society**
*Secretary:* Dave Balcombe, 52 Folly Road, Wymondham, Norfolk, NR18 0QR.
*Web site:* www.norwich.astronomical.society.org.uk
*Meetings:* Seething Observatory, Toad Lane, Thwaite St Mary, Norfolk. Every Friday, 7.30 p.m.

**Nottingham Astronomical Society**
*Secretary:* C. Brennan, 40 Swindon Close, The Vale, Giltbrook, Nottingham NG16 2WD.
*Meetings:* Djanogly City Technology College, Sherwood Rise (B682). 1st and 3rd Thursdays each month, 7.30 p.m.

**Oldham Astronomical Society**
*Secretary:* P. J. Collins, 25 Park Crescent, Chadderton, Oldham.
*Meetings:* Werneth Park Study Centre, Frederick Street, Oldham. Fortnightly, Friday.

**Open University Astronomical Society**
*Secretary:* Dr Andrew Norton, Department of Physics and Astronomy, The Open University, Walton Hall, Milton Keynes MK7 6AA.
*Web site:* www.physics.open.ac.uk/research/astro/a_club.html
*Meetings:* Open University, Milton Keynes. 1st Tuesday of every month, 7.30 p.m.

**Orpington Astronomical Society**
*Secretary:* Dr Ian Carstairs, 38 Brabourne Rise, Beckenham, Kent BR3 2SG.
*Meetings:* High Elms Nature Centre, High Elms Country Park, High Elms Road, Farnborough, Kent. 4th Thursday each month, Sept.–July, 7.30 p.m.

**Papworth Astronomy Club**
*Contact:* Keith Tritton, Magpie Cottage, Fox Street, Great Gransden, Sandy, Bedfordshire SG19 3AA.
*Email:* kpt2@tutor.open.ac.uk
*Meetings:* Bradbury Progression Centre, Church Lane, Papworth Everard, near Huntingdon. 1st Wednesday each month, 7 p.m.

**Peterborough Astronomical Society**
*Secretary:* Sheila Thorpe, 6 Cypress Close, Longthorpe, Peterborough.
*Meetings:* 1st Thursday every month, 7.30 p.m.

**Plymouth Astronomical Society**
*Secretary:* Alan G. Penman, 12 St Maurice View, Plympton, Plymouth, Devon PL7 1FQ.
*Email:* oakmount12@aol.com
*Meetings:* Glynis Kingham Centre, YMCA Annex, Lockyer Street, Plymouth. 2nd Friday each month, 7.30 p.m.

**PONLAF**

*Secretary:* Matthew Hepburn, 6 Court Road, Caterham, Surrey CR3 5RD.

*Meetings:* Room 5, 6th floor, Tower Block, University of North London. Last Friday each month during term time, 6.30 p.m.

**Port Talbot Astronomical Society** (was **Astronomical Society of Wales**)

*Secretary:* Mr J. Hawes, 15 Lodge Drive, Baglan, Port Talbot, West Glamorgan SA12 8UD.

*Meetings:* Port Talbot Arts Centre. 1st Tuesday each month, 7.15 p.m.

**Portsmouth Astronomical Society**

*Secretary:* G. B. Bryant, 81 Ringwood Road, Southsea.

*Meetings:* Monday, fortnightly.

**Preston & District Astronomical Society**

*Secretary:* P. Sloane, 77 Ribby Road, Wrea Green, Kirkham, Preston, Lancs.

*Meetings:* Moor Park (Jeremiah Horrocks) Observatory, Preston. 2nd Wednesday, last Friday each month, 7.30 p.m.

**Reading Astronomical Society**

*Secretary:* Mrs Ruth Sumner, 22 Anson Crescent, Shinfield, Reading RG2 8JT.

*Meetings:* St Peter's Church Hall, Church Road, Earley. 3rd Friday each month, 7 p.m.

**Renfrewshire Astronomical Society**

*Secretary:* Ian Martin, 10 Aitken Road, Hamilton, South Lanarkshire ML3 7YA.

*Web site:* www.renfrewshire-as.co.uk; *Email:* RenfrewAS@aol.com

*Meetings:* Coats Observatory, Oakshaw Street, Paisley. Fridays, 7.30 p.m.

**Rower Astronomical Society**

*Secretary:* Mary Kelly, Knockatore, The Rower, Thomastown, Co. Kilkenny, Eire.

**St Helens Amateur Astronomical Society**

*Secretary:* Carl Dingsdale, 125 Canberra Avenue, Thatto Heath, St Helens, Merseyside WA9 5RT.

*Meetings:* As arranged.

**Salford Astronomical Society**

*Secretary:* Mrs Kath Redford, 2 Albermarle Road, Swinton, Manchester M27 5ST.

*Meetings:* The Observatory, Chaseley Road, Salford. Wednesdays.

**Salisbury Astronomical Society**

*Secretary:* Mrs R. Collins, 3 Fairview Road, Salisbury, Wiltshire, SP1 1JX.

*Meetings:* Glebe Hall, Winterbourne Earls, Salisbury. 1st Tuesday each month.

**Sandbach Astronomical Society**

*Secretary:* Phil Benson, 8 Gawsworth Drive, Sandbach, Cheshire.

*Meetings:* Sandbach School, as arranged.

**Sawtry & District Astronomical Society**

*Secretary:* Brooke Norton, 2 Newton Road, Sawtry, Huntingdon, Cambridgeshire, PE17 5UT.

*Meetings:* Greenfields Cricket Pavilion, Sawtry Fen. Last Friday each month.

**Scarborough & District Astronomical Society**

*Secretary:* Mrs S. Anderson, Basin House Farm, Sawdon, Scarborough, N. Yorks.

*Meetings:* Scarborough Public Library. Last Saturday each month, 7–9 p.m.

**Scottish Astronomers Group**

*Secretary:* Dr Ken Mackay, Hayford House, Cambusbarron, Stirling, FK7 9PR.

*Meetings:* North of Hadrian's Wall, twice yearly.

**Sheffield Astronomical Society**
*Secretary:* Mr Andrew Green, 11 Lyons Street, Ellesmere, Sheffield S4 7QS.
*Web site:* www.saqqara.demon.co.uk/sas/sashome.htm
*Meetings:* Twice monthly at Mayfield Environmental Education Centre, David Lane,
Fulwood, Sheffield S10, 7.30–10 p.m.

**Shetland Astronomical Society**
*Secretary:* Peter Kelly, The Glebe, Fetlar, Shetland, ZE2 9DJ.
*Email:* theglebe@zetnet.co.uk
*Meetings:* Fetlar, Fridays, Oct.–Mar.

**Shropshire Astronomical Society**
*Secretary:* Mrs Jacqui Dodds, 35 Marton Drive, Wellington, Telford, TF1 3HL.
*Web site:* www.astro.cf.ac.uk/sas/sasmain.html; *Email:* jacquidodds@ntlworld.com
*Meetings:* Gateway Arts and Education Centre, Chester Street, Shrewsbury.
Occasional Fridays plus monthly observing meetings, Rodington Village Hall.

**Sidmouth and District Astronomical Society**
*Secretary:* M. Grant, Salters Meadow, Sidmouth, Devon.
*Meetings:* Norman Lockyer Observatory, Salcombe Hill. 1st Monday in each month.

**Skipton & Craven Astronomical Society**
*Contact:* Tony Ireland, 14 Cross Bank, Skipton, North Yorkshire BD23 6AH.
*Email:* scas@beeb.net
*Meetings:* Monthly. Oct.–April. For venue and times contact Mr Ireland.

**Solent Amateur Astronomers**
*Secretary:* Ken Medway, 443 Burgess Road, Swaythling, Southampton SO16 3BL.
*Web site:* www.delscope.demon.co.uk;
*Email:* kenmedway@kenmedway.demon.co.uk
*Meetings:* Room 8, Oaklands, Community School, Fairisle Road, Lordshill,
Southampton. 3rd Tuesday each month, 7.30 p.m.

**Southampton Astronomical Society**
*Secretary:* John Thompson, 4 Heathfield, Hythe, Southampton, SO45 5BJ.
*Web site:* www.home.clara.net/lmhobbs/sas.html
*Email:* John.G.Thompson@Tesco.net
*Meetings:* Conference Room 3, The Civic Centre, Southampton. 2nd Thursday each
month (except Aug.), 7.30 p.m.

**South Downs Astronomical Society**
*Secretary:* J. Green, 46 Central Avenue, Bognor Regis, West Sussex, PO21 5HH.
*Web site:* www.southdowns.org.uk
*Meetings:* Chichester High School for Boys. 1st Friday in each month (except Aug.).

**South-East Essex Astronomical Society**
*Secretary:* C. P. Jones, 29 Buller Road, Laindon, Essex.
*Web site:* www.seeas.dabsol.co.uk/; *Email:* cpj@cix.co.uk
*Meetings:* Lecture Theatre, Central Library, Victoria Avenue, Southend-on-Sea.
Generally 1st Thursday in month, Sept.–May, 7.30 p.m.

**South-East Kent Astronomical Society**
*Secretary:* Andrew McCarthy, 25 St Paul's Way, Sandgate, near Folkestone, Kent,
CT20 3NT.
*Meetings:* Monthly.

**South Lincolnshire Astronomical & Geophysical Society**
 *Secretary:* Ian Farley, 12 West Road, Bourne, Lincolnshire, PE10 9PS.
 *Meetings:* Adult Education Study Centre, Pinchbeck. 3rd Wednesday each month,
 7.30 p.m.
**Southport Astronomical Society**
 *Secretary:* Patrick Brannon, Willow Cottage, 90 Jacksmere Lane, Scarisbrick,
 Ormskirk, Lancashire, L40 9RS.
 *Meetings:* Monthly Sept.–May, plus observing sessions.
**Southport, Ormskirk and District Astronomical Society**
 *Secretary:* J. T. Harrison, 92 Cottage Lane, Ormskirk, Lancs L39 3NJ.
 *Meetings:* Saturday evenings, monthly as arranged.
**South Shields Astronomical Society**
 *Secretary:* c/o South Tyneside College, St George's Avenue, South Shields.
 *Meetings:* Marine and Technical College. Each Thursday, 7.30 p.m.
**South Somerset Astronomical Society**
 *Secretary:* G. McNelly, 11 Laxton Close, Taunton, Somerset.
 *Meetings:* Victoria Inn, Skittle Alley, East Reach, Taunton, Somerset. Last Saturday
 each month, 7.30 p.m.
**South-West Hertfordshire Astronomical Society**
 *Secretary:* Tom Walsh, 'Finches', Coleshill Lane, Winchmore Hill, Amersham,
 Buckinghamshire HP7 0NP.
 *Meetings:* Rickmansworth. Last Friday each month, Sept.–May.
**Stafford and District Astronomical Society**
 *Secretary:* Miss L. Hodkinson, 6 Elm Walk, Penkridge, Staffordshire, ST19 5NL.
 *Meetings:* Weston Road High School, Stafford. Every 3rd Thursday, Sept.–May,
 7.15 p.m.
**Stirling Astronomical Society**
 *Secretary:* Hamish MacPhee, 10 Causewayhead Road, Stirling FK9 5ER.
 *Meetings:* Smith Museum & Art Gallery, Dumbarton Road, Stirling. 2nd Friday each
 month, 7.30 p.m.
**Stoke-on-Trent Astronomical Society**
 *Secretary:* M. Pace, Sundale, Dunnocksfold, Alsager, Stoke-on-Trent.
 *Meetings:* Cartwright House, Broad Street, Hanley. Monthly.
**Stratford-upon-Avon Astronomical Society**
 *Secretary:* Robin Swinbourne, 18 Old Milverton, Leamington Spa, Warwickshire,
 CV32 6SA.
 *Meetings:* Tiddington Home Guard Club. 4th Tuesday each month, 7.30 p.m.
**Sunderland Astronomical Society**
 *Contact:* Don Simpson, 78 Stratford Avenue, Grangetown, Sunderland SR2 8RZ.
 *Meetings:* Friends Meeting House, Roker. 1st, 2nd and 3rd Sundays each month.
**Sussex Astronomical Society**
 *Secretary:* Mrs C. G. Sutton, 75 Vale Road, Portslade, Sussex.
 *Meetings:* English Language Centre, Third Avenue, Hove. Every Wednesday,
 7.30–9.30 p.m., Sept.–May.
**Swansea Astronomical Society**
 *Secretary:* Dr Michael Morales, 238 Heol Dulais, Birch Grove, Swansea SA7 9LH.
 *Web site:* www.crysania.co.uk/sas/astro/star
 *Meetings:* Lecture Room C, Science Tower, University of Swansea. 2nd and 4th
 Thursday each month from September to June, 7 p.m.

**Tavistock Astronomical Society**
*Secretary:* Mrs Ellie Coombes, Rosemount, Under Road, Gunnislake, Cornwall
PL18 9JL.
*Meetings:* Science Laboratory, Kelly College, Tavistock. 1st Wednesday each month,
7.30 p.m.

**Thames Valley Astronomical Group**
*Secretary:* K. J. Pallet, 82a Tennyson Street, South Lambeth, London SW8 3TH.
*Meetings:* As arranged.

**Thanet Amateur Astronomical Society**
*Secretary:* P. F. Jordan, 85 Crescent Road, Ramsgate.
*Meetings:* Hilderstone House, Broadstairs, Kent. Monthly.

**Torbay Astronomical Society**
*Secretary:* Tim Moffat, 31 Netley Road, Newton Abbot, Devon, TQ12 2LL.
*Meetings:* Torquay Boys' Grammar School, 1st Thursday in month; and Town Hall,
Torquay, 3rd Thursday in month, Oct.–May, 7.30 p.m.

**Tullamore Astronomical Society**
*Secretary:* Tom Walsh, 25 Harbour Walk, Tullamore, Co. Offaly, Eire.
*Web site:* www.iol.ie/seanmck/tas.htm; *Email:* tcwalsh25@yahoo.co.uk
*Meetings:* Order of Malta Lecture Hall, Tanyard, Tullamore, Co. Offaly, Eire.
Mondays at 8 p.m., every fortnight.

**Tyrone Astronomical Society**
*Secretary:* John Ryan, 105 Coolnafranky Park, Cookstown, Co. Tyrone.
*Meetings:* Contact Secretary.

**Usk Astronomical Society**
*Secretary:* Bob Wright, 'Llwyn Celyn', 75 Woodland Road, Croesyceiliog, Cwmbran,
NP44 2OX.
*Meetings:* Usk Community Education Centre, Maryport Street, Usk. Every Thursday
during school term, 7 p.m.

**Vectis Astronomical Society**
*Secretary:* Rosemary Pears, 1 Rockmount Cottages, Undercliff Drive, St Lawrence,
Ventnor, Isle of Wight PO38 1XG.
*Web site:* www.wightskies.fsnet.co.uk/main.html
*Email:* may@tatemma.freeserve.co.uk
*Meetings:* Lord Louis Library Meeting Room, Newport. 4th Friday each month
except Dec., 7.30 p.m.

**Vigo Astronomical Society**
*Secretary:* Robert Wilson, 43 Admers Wood, Vigo Village, Meopham, Kent
DA13 0SP.
*Meetings:* Vigo Village Hall. As arranged.

**Walsall Astronomical Society**
*Secretary:* Bob Cleverley, 40 Mayfield Road, Sutton Coldfield, B74 3PZ.
*Meetings:* Freetrade Inn, Wood Lane, Pelsall North Common. Every Thursday.

**Wellingborough District Astronomical Society**
*Secretary:* S. M. Williams, 120 Brickhill Road, Wellingborough, Northants.
*Meetings:* Gloucester Hall, Church Street, Wellingborough. 2nd Wednesday each
month, 7.30 p.m.

**Wessex Astronomical Society**
*Secretary:* Leslie Fry, 14 Hanhum Road, Corfe Mullen, Dorset.
*Meetings:* Allendale Centre, Wimborne, Dorset. 1st Tuesday of each month.

**West Cornwall Astronomical Society**
*Secretary:* Dr R. Waddling, The Pines, Pennance Road, Falmouth, Cornwall
TR11 4ED.
*Meetings:* Helston Football Club, 3rd Thursday each month, and St Michalls Hotel,
1st Wednesday each month, 7.30 p.m.

**West of London Astronomical Society**
*Secretary:* Duncan Radbourne, 28 Tavistock Road, Edgware, Middlesex HA8 6DA.
*Web site:* www.wocas.org.uk
*Meetings:* Monthly, alternately in Uxbridge and North Harrow. 2nd Monday in
month, except Aug.

**West Midlands Astronomical Association**
*Secretary:* Miss S. Bundy, 93 Greenridge Road, Handsworth Wood, Birmingham.
*Meetings:* Dr Johnson House, Bull Street, Birmingham. As arranged.

**West Yorkshire Astronomical Society**
*Secretary:* Pete Lunn, 21 Crawford Drive, Wakefield, West Yorkshire.
*Meetings:* Rosse Observatory, Carleton Community Centre, Carleton Road,
Pontefract. Each Tuesday, 7.15 p.m.

**Whitby and District Astronomical Society**
*Secretary:* Rosemary Bowman, The Cottage, Larpool Drive, Whitby, North
Yorkshire, YO22 4ND.
*Meetings:* Whitby Mission, Seafarer's Centre, Haggersgate, Whitby. 1st Tuesday of
the month, 7.30 p.m.

**Whittington Astronomical Society**
*Secretary:* Peter Williamson, The Observatory, Top Street, Whittington, Shropshire.
*Meetings:* The Observatory. Every month.

**Wiltshire Astronomical Society**
*Secretary:* Simon Barnes, 25 Woodcombe, Melksham, Wilts SN12 6HA.
*Meetings:* St Andrews Church Hall, Church Lane, off Forest Road, Melksham, Wilts.

**Wolverhampton Astronomical Society**
*Secretary:* Mr M. Bryce, Iona, 16 Yellowhammer Court, Kidderminster,
Worcestershire, DY10 4RR.
*Web site:* www.wolvas.org.uk; *Email:* michaelbryce@wolvas.org.uk
*Meetings:* Beckminster Methodist Church Hall, Birches Barn Road, Wolverhampton.
Alternate Mondays, Sept.–Apr., extra dates in summer, 7.30 p.m.

**Worcester Astronomical Society**
*Secretary:* Mr S. Bateman, 12 Bozward Street, Worcester WR2 5DE.
*Meetings:* Room 117, Worcester College of Higher Education, Henwick Grove,
Worcester. 2nd Thursday each month, 8 p.m.

**Worthing Astronomical Society**
*Contact:* G. Boots, 101 Ardingly Drive, Worthing, West Sussex, BN12 4TW.
*Web site:* www.worthingastro.freeserve.co.uk
*Email:* gboots@observatory99.freeserve.co.uk
*Meetings:* Heene Church Rooms, Heene Road, Worthing. 1st Wednesday each
month (except Aug.), 7.30 p.m.

**Wycombe Astronomical Society**
*Secretary:* Mr P. Treherne, 34 Honeysuckle Road, Widmer End, High Wycombe,
Buckinghamshire, HP15 6BW.
*Meetings:* Woodrow High House, Amersham. 3rd Wednesday each month, 7.45 p.m.

**The York Astronomical Society**
 *Contact:* Hazel Collett, Public Relations Officer
 *Telephone:* 07944 751277
 *Web site:* www.yorkastro.freeserve.co.uk; *Email:* info@yorkastro.co.uk
 *Meetings:* The Knavesmire Room, York Priory Street Centre, Priory Street, York.
 1st and 3rd Friday of each month (except Aug.), 8 p.m.

Any society wishing to be included in this list of local societies or to update details, including any web-site addresses, is invited to write to the Editor (c/o Pan Macmillan, 20 New Wharf Road, London N1 9RR), so that the relevant information may be included in the next edition of the *Yearbook*.

---

The William Herschel Society maintains the museum established at 19 New King Street, Bath BA1 2BL – the only surviving Herschel House. It also undertakes activities of various kinds. New members would be welcome; those interested are asked to contact the Membership Secretary at the museum.

---

The South Downs Planetarium (Kingsham Farm, Kingsham Road, Chichester, West Sussex PO19 8RP) is now fully operational. For further information, visit www.southdowns.org.uk/sdpt or telephone (01243) 774400